T0199367

GEOGRAPHIC INFORMATION

Value, Pricing, Production, and Consumption

GEOGRAPHIC INFORMATION

Value, Pricing, Production, and Consumption

Roger A. Longhorn

Michael Blakemore

CRC Press
Taylor & Francis Group
Boca Raton London New York

CRC Press is an imprint of the
Taylor & Francis Group, an **informa** business

CRC Press
Taylor & Francis Group
6000 Broken Sound Parkway NW, Suite 300
Boca Raton, FL 33487-2742

First issued in paperback 2020

© 2008 by Taylor & Francis Group, LLC
CRC Press is an imprint of Taylor & Francis Group, an Informa business

No claim to original U.S. Government works

ISBN-13: 978-0-367-57758-2 (pbk)
ISBN-13: 978-0-8493-3414-6 (hbk)

Library of Congress Cataloging-in-Publication Data

Longhorn, Roger A.
   Geographic information : value, pricing, production, and consumption / Roger A. Longhorn and Michael Blakemore.
      p. cm.
   Includes bibliographical references and index.
   ISBN 978-0-8493-3414-6 (alk. paper)
   1. Geographical information systems--Economic aspects. I. Blakemore, M. J. II. Title.

G70.212.L656 2008
910.285--dc22   2007025544

Visit the Taylor & Francis Web site at
http://www.taylorandfrancis.com

and the CRC Press Web site at
http://www.crcpress.com

Contents

Preface

The original concept for this book sprang from the reception we received to a co-authored article that appeared in the March 2004 issue of the online peer-reviewed e-magazine *Journal of Digital Information* (Longhorn and Blakemore, 2004). In that paper, we challenged the dogma — the almost religious fervor — evident in the opposing viewpoints that characterized the debate on charging for public sector information (PSI), i.e., fee or free. This polarization seemed especially vehement in relation to geographic information (GI), which is claimed to be highly valuable and expensive to collect and maintain while inexpensive to disseminate. The paper widened the debate to include the economic reality of the information market, in both the private and public sectors, and the impact of diverse public information policy cultures on pricing, charging, access, and exploitation of GI. The current text represents the authors' attempt to expand on that initial paper following a further three years of research.

Following a scene-setting chapter drawing partly from the original March 2004 paper, Chapter 2 looks at the many ways that information can be valued, from the theoretical viewpoint of value theory and value chains in an information market setting, to specific attributes of GI that have positive — and negative — impacts on its value to different users. One conclusion reached is that it is often not possible to assign a single, constant value to specific GI due to the number of variables inherent in how that GI is produced and used. Be forewarned that this chapter does not contain a formal economic analysis of the value of GI for the simple reason that a complete text on that topic would be required to do it justice. Also, we have found that, in practice, the decision makers who judge the value of GI and set the pricing and charging policies relating to GI seem not to pay too close attention to the economic theories now extant.

Chapter 3 focuses on collecting, disseminating, and using GI in the widest sense of the term *business*; i.e., not specifically relating to commercial enterprises, but to any organization that must collect, process, maintain, disseminate, and use GI. The key premise is posed in the chapter subtitle: no such thing as a free lunch. Recognize that all information has a cost, in fact, a range of costs, associated with it, and someone has to pay these somewhere, somehow, sometime. We try to bring some objectivity to the charging and cost recovery debate relating to public sector GI by relating the reality of developments in the information market with the expectations of different stakeholders who collectively comprise the GI producer and user communities.

Chapter 4 looks at pricing of information, from basic theory to pricing models applied by producers of GI in the public sector. Traditional price discrimination theories are extended to include the free lunch (zero-degree pricing) referred to in Chapter 3. Other pricing issues are exposed, such as the impact of time delays in acquiring information, quality, revenue sharing options, product differentiation, and uncertainty. The chapter concludes with a look at the dynamic, changing relationships between information producers and users, and the impact they have on the information content industry generally and GI specifically.

Chapter 5 introduces a more global look at GI, beginning with its claimed ubiquity — a myth yet to be proved or disproved. What is the impact of focusing primarily on a location attribute, whose presence, among many other important attributes for a piece of information, leads to the label of "geographic information," when that label may have value only to those who work in the GI industry? Social-technical aspects of GI and geographic information systems (GISs) are examined via real-world examples, followed by a look at GI and PSI governance in regard to spatial data infrastructures (SDIs). GI globalization, repurposing of GI, and the impact of information overload round out the chapter.

Chapter 6 examines the SDI phenomenon from strategy to policy to implementation, providing a review of key SDI policy trends globally, and access and pricing policies more specifically. Uncertainties facing decision makers who must find and approve funding for creating SDIs, e.g., suitable cost–benefit analysis methodologies, are explored, including an overview of such studies spanning more than 15 years. The chapter concludes with a recommendation on the value of cost–benefit methodologies in various scenarios.

Chapter 7 brings the book to a conclusion with a summary of the authors' thoughts on the main topics presented so far and prospects for the future.

Those hoping to find here a new academic treatment on GI valuation, information economics, pricing, and charging will be disappointed. Rather, we have adopted a style and format that further widens the debate on these important issues. New viewpoints are presented, drawing parallels from other sectors of the information market, as well as noninformation markets. Our goal with this book is to stimulate the debate, while defusing some of the current highly polarized fee or free dogma relating to charging for PSI, especially in relation to GI. More stakeholders need to join this debate, with open and questioning minds, especially the decision makers responsible for creating SDIs at local, regional, national, and global levels, in both developing and developed countries.

No one doubts the value of geographic information, even if we cannot always attach an objective, monetary cost–benefit or positive return on investment to its collection, maintenance, and use. The information world is constantly changing, continually evolving, and numerous new models of information management and use are appearing in both the commercial and public sectors. We should not let dogma act as a barrier to the most effective use of GI, regardless of where it originates.

References

1. Longhorn, R. and M. Blakemore. 2004. Re-visiting the valuing and pricing of digital geographic information. *Journal of Digital Information*, 4: 1–27. http://jodi.tamu.edu/Articles/v04/i02/Longhorn/ (accessed April 6, 2007).

Acknowledgments

There is a diverse community from around the world that has advised, influenced, and critiqued our work over several decades. It seems pretentious, even egoistic, to provide a huge list, and we would worry about missing people. Nevertheless, there are a few individuals who have been significantly influential. We both believe that Professor David Rhind remains one of the most critically constructive researchers in the world of geographic information, which he achieves while maintaining humor, friendship, and an open mind. That is something to which we should all aspire. David's depth and breadth of experience, spanning decades in the GI industry, both in academia and as former head of Ordnance Survey of Great Britain, provide him with a unique view on developments and key issues in this industry.

Helpful comments were also provided on certain chapters by the late Mike Clark, who served on the U.K. government's Advisory Panel for Public Sector Information (APPSI). Mike shared with the authors a desire to defuse the dogmatic approach to questions dealing with charging for public sector information, without taking sides on the issue or supporting arguments that are not backed by evidence. If more people in the GI community and industry would adopt a similar attitude, we could all have more sensible discussions on these issues in the future.

Mike Blakemore especially thanks the one person who has been a source of advice throughout his career, Professor Peter Lloyd, who rescued Mike at an early stage from the clutches of abstract academic introversion, gave him a business focus, and taught him the value of information. Roger Longhorn personally thanks all those who participated in hundreds of online discussions on the issues covered in this book, via e-mail, or at innumerable GI/GIS events he has attended globally over the past 15 years.

We acknowledge with gratitude the patience shown to us by the team at Taylor & Francis. We must be very frustrating authors to deal with, and our timelines occasionally must have seemed geological, not geographical, to them. A similar experience must also have been felt by Flo (Longhorn) and Irene (Blakemore), who both maintained a suitable puzzled, yet tolerant, attitude as we continued to change the writing as we read yet another interesting reference. This book is dedicated to them. Heavens only knows what they will think when this is finally in print!

About the Authors

Roger Longhorn is co-director of Info-Dynamics Research Associates Ltd., where he conducts research into a range of information market and society issues, focusing especially on national and regional spatial information infrastructures. He has been involved in the ICT industry since leaving MIT in 1976, as a commercial MIS developer, expert in information services for the European Commission's DG Information Society and DG Enterprise, and independent consultant on a range of information market and infrastructure projects for national governments. Roger assisted in developing the first strategy for geospatial information infrastructure in Europe (GI2000) and continues to be active in implementation of the Infrastructure for Spatial Information in Europe (INSPIRE) directive. An expert in EU and global information policy for geospatial data, Roger has served on EC and European Space Agency (GMES) review and strategy advisory panels for GIS-related projects. He specializes in ICT policy analysis and since 2000 has conducted extensive independent research into spatial data policy and strategy, including cost–benefit analysis methodologies for information infrastructures.

Roger is steering committee leader for MOTIIVE, a project funded by the EU to implement rules of the INSPIRE Directive. He has advised the Irish, Spanish regional (Catalunya), and Egyptian governments on SDI strategy, policy, and architectures. He is information policy advisor to EUCC — the Coastal Union (Leiden, the Netherlands) and a member of the Commission on Coastal Systems of the International Geographical Union (IGU) — as well as co-chair of the Legal & Economic Working Group of the Global Spatial Data Infrastructure (GSDI) Association and a member of EUROGI, the European umbrella organization for GI. Roger also provides external expert services to various directorates general of the European Commission in the realm of GI and GIS projects and programs.

Michael J. Blakemore is emeritus professor of geography at the University of Durham, following an academic career in geography spanning nearly 30 years. Mike is a geographer whose activities encompass information science, history, official statistics, e-government and e-society strategy and policy, and international development. He has advised national and international government policy, has taught widely on geographic information

(GI), government, and society, and is active in research that develops new debates regarding access to geographic information, its value, cost, and pricing, and its wider uses within the processes of globalization. He developed successful geostatistical data businesses within the university requirements, with extensive expertise gained in pricing and dissemination strategy, organizational behavior, customer development, metadata, and training. Mike is an experienced international speaker, having given over 100 conference and seminar presentations around the world, in particular to large international audiences and senior management and decision makers. He is author and co-author of over 140 publications, and teaches widely on information policy, strategy, and society.

Mike is co-director of Info-Dynamics Research Associates Ltd., a U.K. consultancy that conducts research into a range of information market and society issues, including e-government and national and regional information infrastructures, and provides external expert services to various directorates general of the European Commission.

chapter one

Introduction

As the title suggests, this book is first and foremost about geographic information (GI) and how society assigns different values to GI* and makes it available for exploitation; especially the for-free or for-a-fee debate surrounding GI produced in, by, or for the public sector — so-called public sector GI (PSGI). Various studies from developed nations around the world report that GI plays an important role in underpinning economies, delivering more efficient government, enhancing quality of life for citizens, improving business efficiency, and generating new business and employment opportunities. Such benefits would indicate that GI should be used as widely as possible (Baltimore County, 2001; Booz Allen Hamilton, 2005; CIE, 2000; Craglia and INSPIRE FDS Working Group, 2003; Halsing et al., 2004; Hardwick and Fox, 1999; Montgomery County Council, 1999; OXERA, 1999; PIRA, 2000; Price Waterhouse, 1995; Werschler and Rancourt, 2005). Much GI is collected by local and national government for specific purposes, either legally mandated or required to improve operational efficiency. How such public sector information (PSI) is made more widely available for other uses and to other users, at what price and with or without restrictions on reuse, has created heated debate and led to the adoption of diverse PSI charging regimes in different countries (Longhorn and Blakemore, 2004). The overall goal of this book is to address the apparent dogma inherent in the often bipolar viewpoints surrounding the PSGI pricing and charging debate, taking into consideration the differing values of GI, the role of GI and PSGI in society generally, and the impact of the debate on evolving spatial data infrastructures (SDIs) from the perspectives of economic reality and diverse public information policy cultures.

The authors have commercial and academic experience with data access, exploitation, and pricing issues and policies, in both the private and public sectors, spanning nearly two decades. Our combined experience led to the belief that public sector information debates, which began more than 20 years ago, often fail to progress beyond entrenched positions based on ideology and emotion; sometimes based on myths about PSI that are perpetuated even today. The value of geographic information often is misunderstood or naively assigned from individual viewpoints that do not encompass the whole range of issues surrounding the production, maintenance, distribution, and consumption of GI — in other words the whole GI life cycle — and

---

* The acronym *GI* as used in this chapter and throughout the book should be taken as synonymous with terms such as *geographic data, geospatial information, spatial information, geospatial* or *spatial data,* or similar terms now widely used in much of the literature.

for different types of GI. Existing access, pricing, charging, and exploitation policies are often complicated, rife with contradictions and inconsistencies across government, even within single states, and sometimes even within single agencies.

Agreeing on a common definition for geographic information is the first step in entering the debate, or at least understanding the nuances that different definitions may bring to the debate, depending upon the definers' viewpoint. The most simplistic definition of GI — all information with a location attribute — instantly spans a huge realm of data, from addresses, to physical and nonphysical boundaries, to discernible features of the natural and built environment, in two dimensions, in three dimensions, and over time. Yet there are considerable, practical differences between, for example, GI defining real-world features, such as a road network, river, or coastline, whether represented by vectors or raster images, and GI consisting solely of a person's address, assigned as only one attribute to a plethora of other information describing his or her medical condition, financial or employment status, or educational achievements. For instance, the type of GI collected dramatically impacts on cost of collection and maintenance, on distribution and use, and on legal, commercial, and privacy issues. This introductory chapter explores these definitions in more detail and the impact that different perceived definitions can have on other parts of the PSGI debate. The next section of this chapter presents some of the definitions currently in use or adopted over time, and proposes a more comprehensive definition for the twenty-first century.

1.1   What is geographic information?

One problem with current definitions of geographic information is that they appear to be either too general or too specific, too simplistic or too technically (GIS) oriented, or they vary in other subtle and nonsubtle ways, depending upon what issues relating to GI are under discussion, i.e., collecting, storing, using, valuing, charging, etc. Experience from numerous public debates, and as evidenced in SDI framework specifications at national, regional, and global levels, including from official standards bodies, indicates that there is not a single agreed-upon definition for the term *geographic information*. Rather, a range of terms are in use, often interchangeably, but with different meanings to different communities. Definitions from national, regional, and global bodies include:

- "Spatial information (also known as geographic information) ... any information that can be geographically referenced, i.e. describing a location or any information that can be linked to a location" (ANZLIC, 2006).
- "Spatial data" is "any data with a direct or indirect reference to a specific location or geographical area" (EU, 2007).

- "Geographic information" is "information concerning phenomena implicitly or explicitly associated with a location relative to the Earth" (ISO, 2002; CEN, 1998).

The problem is that such general definitions are of little practical use in assigning or assessing specific values for a specific type or instance of use of information that has a location attribute among many other attributes, or indeed to the location attribute itself. We say this because these definitions do not discern between the location attribute compared to the many other attributes that may exist within, and comprise, a specific piece of information as a whole. When we get to the value and pricing issues relating to GI, these are important considerations.

Definitions used or endorsed by the U.K. Association for Geographic Information (AGI) also show subtle changes over time. In 1991, GI was defined as "information which can be related to a location (defined in terms of point, area or volume) on the Earth, particularly information on natural phenomena, cultural and human resources. A special case of spatial information" (AGI, 1991; Maguire et al., 1991). In that same publication, *spatial information* is defined as "information which includes a reference to a two or three dimensional position in space as one of its attributes." By 1996, other AGI publications were including definitions such as information that "includes any data about areas, objects, statistics or records which include a spatial reference (e.g. a grid reference or postcode)." The shift in emphasis is from the earliest definitions for GI as any information that can be related to a location, to acknowledgment, in the spatial data definition, that the location is but one attribute, to the later definition that explicitly moves on from the "natural or man-made phenomena" class to information instances that include "objects, statistics, or records" that simply have a locational reference attribute, which may itself need georeferencing to a location on earth, e.g., a postcode.

From 1999, *spatial data* continues to be defined as "any information about the location and shape of, and relationships among, geographic features. This includes remotely sensed data as well as map data" (AGI, 1999). The U.S. Federal Geographic Data Committee defines *geospatial data* as "information that identifies the geographic location and characteristics of natural or constructed features and boundaries on the earth. This information may be derived from, among other things, remote sensing, mapping, and surveying technologies" (FGDC, 2007). These definitions now add a new dimension, i.e., how the data are gathered, presented, or analyzed (e.g., remote sensing, surveying), while at the same time reverting back to the geographic features theme without further reference to other attributes, such as those recognized in the next example from 1987.

Some definitions attempt to be more explicit, by offering illustrative examples, such as this definition from the 1987 Chorley Report for the U.K. government, in which *geographic information* is:

> Information which can be related to specific locations
> on the Earth ... including the distribution of natural
> resources, the incidence of pollutants, descriptions of
> infrastructure such as buildings, utility and transport
> services, patterns of land use and the health, wealth,
> employment, housing and voting habits of people.
> (DOE, U.K., 1987)

Here we see incorporated examples of numerous spatial themes other than physical environment or topography, touching on demographic data, e.g., "health, wealth, employment ... voting habits."

Does the U.K. National Grid constitute geographic information? This grid system is an artificial construct that overlies the U.K. landmass, extending slightly seaward. The grid is used to assign position or location to other forms of information, whether natural or man-made features or administrative boundaries of various sorts, ranging from electoral ward and county boundaries to river catchment areas and addresses. In fact, the first and highest priority data theme in the pan-European SDI directive, INSPIRE, is "1. Coordinate reference systems — Systems for uniquely referencing spatial information in space as a set of coordinates (x, y, z) and/or latitude and longitude and height, based on a geodetic horizontal and vertical datum" (EU 2007, Annex 1). But how can one attach a value to an entire coordinate system, financial or otherwise?

Another interesting variation is provided in the dictionary of GIS technology from ESRI, one of the world's largest GIS vendors, which defines *geographic data* (ESRI, 2001) as "information about geographic features, including their locations, shapes and descriptions" — but has no separate definition for *geographic information*. Spatial data is defined as "1. Information about the locations and shapes of geographic features, and the relationships between them; usually stored as coordinates and topology. 2. Any data that can be mapped." Here, definitions for GI relate more to the GIS technology, with references to "shapes" and "coordinates and topology," all data for mapping, an important function of GIS tool sets. Along similar lines, Blinn and co-authors (2007) of an online GIS glossary on a University of Minnesota website define *geographic data* as "data that convey the locations and descriptions of geographic features" and *spatial data* as "data pertaining to the location, shape, and relationships among geographical features. These can be classified and stored as point, line, area, polygon, grid cell, or object" (Blinn et al., 2007). Again, we have definitions that focus on location attributes and geometries, specifically for geographical features. They do not mention other nonlocational aspects or attributes of a piece of information, such as a tax record, health record, or value of a house. Interestingly, these and other such glossaries from within the GIS community seldom describe or define geographic *information*, but rather concentrate on the *data* — geographic or spatial — that IT systems are designed to manage and process.

ESRI's later online GIS dictionary support site (ESRI, 2007) defines *geographic data* as "information describing the location and attributes of things, including their shapes and representation. Geographic data is the composite of spatial data and attribute data." Another online glossary defines *geographic data* as "the locations and descriptions of geographic features. The composite of spatial data and descriptive data" (GIS Development, 2007). These now explicitly recognize the joining of spatial (location) data with other attributes. However, ESRI's 2007 online definition for spatial data remains the same as in the 2001 printed dictionary mentioned above. Most current online dictionaries continue to ignore the term *information* altogether, preferring to focus on *data* that is manipulated by GIS or IT tools.

Geographic information is often characterized by high data volumes per product, but not necessarily so, e.g., high-resolution images of tens of megabytes each or large databases vs. single land registry boundaries or addresses or point data on locations of specific features of interest. Yet size, quality, or even number of records in a data set may not relate to value, as we shall see in Chapter 2, since value depends upon so many other factors.

Look at the 34 data themes in the EU's INSPIRE directive in Table 1.1 and see for how many of these themes the location attribute is the prime attribute of value vs. representing only one of many other important attribute values. The difference is mainly between location of known features and artificial (legal, administrative, geodemographic) boundaries, addresses, etc. These are key, basic GI — mainly topographic, but not totally. Then all other data that are claimed to be GI are actually scientific, commercial, or for governance, and location is only one attribute that has value solely when spatial analysis is required, not otherwise.

Having exposed several different meanings for the term *geographic information* and its many cousins, we propose that a comprehensive meaning for the twenty-first century would read something like that shown in the following box. This draws on prior definitions of types of data, combined with potential uses of the data, and removes limitations relating specifically to physical environment or any one type of coordinate system, since many are in use today; e.g., different types of national or thematic grids, two- and three-dimensional meshes, lat-long, lat-long and depth or height, etc.

Geographic information is a composite of spatial data and attribute data describing the location and attributes of things (objects, features, events, physical or legal boundaries, volumes, etc.), including the shapes and representations of such things in suitable two-dimensional, three-dimensional, or four-dimensional (x, y, z, time) reference systems (e.g., a grid reference, coordinate system reference, address, postcode, etc.) in such a way as to permit spatial (place-based) analysis of the relationships between

Table 1.1   INSPIRE Directive Spatial Data Themes

Highest Priority Data Themes in INSPIRE Directive
Coordinate reference system (x, y, z or lat/long)
Geographical grid systems (harmonized multiresolution grid)
Geographical names
Administrative units (local, regional, and national boundaries)
Addresses
Cadastral parcels
Transport networks (road, rail, air, water, and links between networks)
Hydrography (including marine areas, all water bodies, river basins, etc.)
Protected sites (designated by national, EU, or international legislation)
Second Highest Priority Data Themes in INSPIRE Directive
Elevation (land, ice, and ocean surfaces; terrestrial elevation, bathymetry, shoreline)
Land cover (physical and biological)
Orthoimagery (georeferenced image data)
Geology (including bedrock, aquifers, and geomorphology)
Third Highest Priority Data Themes in INSPIRE Directive
Statistical units (for dissemination or use of statistical data)
Buildings (geographical location of buildings)
Soil (and subsoil characteristics)
Land use (e.g., residential, industrial, commercial)
Human health and safety (see full description in annex)
Utility and governmental services (sewage, waste management, energy, etc.)
Environmental monitoring facilities (emissions, ecosystem parameters)
Production and industrial facilities (water abstraction, mining, storage sites)
Agricultural and aquacultural facilities
Population distribution — demography
Area management/restrictions/regulation zones/reporting units
Natural risk zones (e.g., atmospheric, hydrologic, seismic, volcanic, wildfire)
Atmospheric conditions
Meteorological geographical features (weather conditions, measurements)
Oceanographic geographical features (currents, salinity, wave heights, etc.)
Sea regions (physical conditions of seas and saline water bodies)
Biogeographical regions (areas with homogeneous ecological conditions)
Habitats and biotopes (geographical areas for specific ecological conditions)
Species distribution (geographical boundaries for animal and plant species)
Energy resources (hydrocarbons, hydropower, bioenergy, solar, wind, etc.)
Mineral resources (metal ores, industrial minerals depth/height, etc.)

Source: Directive 2007/2/EC of the European Parliament and of the Council of 14 March 2007 establishing an Infrastructure for Spatial Information in the European Community (INSPIRE), Annexes I, II and III, April, 25 2007. *Official Journal of the European Union* (Luxembourg).

and among the things so described, including their different attributes.

Corollary to the definition: The format of portrayal or use does not exclude one type of information from being considered geographic. Geographic information may exist in any number of forms and formats; e.g., an aerial image of a house or street, showing its relationship to other houses and streets, qualifies as geographic information, just as the vectors describing the boundaries of the house or centerline of the road in an x-y coordinate system would.

This, then, is the rather long definition we have in mind whenever we use the term *geographic information* throughout the rest of the book, especially in relation to discussions on value, which underpins the subsequent discussions relating to pricing, charging, and cost–benefit that appear in later chapters.

1.2   Is geographic information unique?

Having now imposed our own definition of geographic information on the reader, we turn to the question of whether GI is somehow unique in the information arena and especially in the information market. The question needs to be asked because, once again, the answer impacts directly on value, and later on pricing, charging, and other issues impacting on the main fee-or-free debate in later chapters.

First, one would say that GI is obviously unique in that it has a location attribute that is absent from other data. The location component or attribute is what permits us to analyze various data sets spatially. Yet that is usually only one important or valuable way that most data sets are used or analyzed, and here is where our new, more comprehensive definition comes into play. A tax record for an individual or a business is a very important piece of information to society, i.e., to government and to the individual or business involved, and it contains many valuable attributes, not least, perhaps, the tax due from — or owing back to — the taxable entity, in which year, due or payable by what date, with or without interest at what rate, penalties due or owing, etc. This tax record is a single piece of composite information typically referenced to a single taxable entity. And one attribute will almost certainly be an address that can be georeferenced to a national grid or other coordinate system to permit some form of spatial analysis or spatial portrayal, if necessary. Yet do we classify the whole tax record as geographic information or only the address attribute and its possible georeferencing characteristics or portrayal? To us, this is the heart of the test and debate over the uniqueness of GI compared to other forms of information.

Looking objectively at this tax record, the information contained therein can be analyzed in many different ways, only one of which is spatially. For example, a time-based analysis could be useful to see how the entity taxed or the government collecting the tax benefits or suffers from the taxation of this entity, historically, today, and forecast into the future regarding, for example, tax rates, taxable bands, payment terms, etc. Performed across a large number of such entities, the government might ask if the tax rates or bands need adjusting or special discounts are needed for certain types of businesses or individuals, or should be removed. All these questions could be examined from using different attributes within the composite data that comprise the tax record. Only if that very important analysis were then extended to look at the impact on specific regions of the country (assuming that taxes are apportioned regionally) would the single spatial attribute begin to have value. This is a fairly naïve example, but it makes the point. To call this tax record or a national collection of such tax records geographic information seems a bit odd, certainly to those operating in the financial community, even though that community is waking today to the value of spatial analysis. The same view would apply to many types of information that the GI community insists on calling geographic information.

Is GI unique because of the high cost of collecting the data, or maintaining or processing it? Well, probably not in relation to other types of information that are equally important to society, such as the whole scientific, technical, and medical (STM) information market that predates the focus on GI by many years. In fact, during the early years of the European Union's information market promotion programs at DG Information Society in the mid-1990s, the values of the GI and GIS market sectors were found to be relatively small (460 to 750 million euro per annum in 1997) compared to almost all other types of information, including STM (U.S.$2.5 billion and growing fast in 1997), and miniscule compared to media content (376 billion euro — 5% of EU gross domestic product, or GDP, in 1998) (Prodger and Sutherland, 1997; Waltham, 2002; Garribba, 1999). This was one of the reasons that it took so many years for the European Commission to find its local champions to drive the pan-European SDI initiative to final fruition with the INSPIRE directive in April 2007 —15 years after the information market programs began to allocate at least some portion of the program budget to GI and GIS. Yes, it may be costly to collect and maintain current, high-quality data on the transport network or natural environment, but it is equally costly, if not more so, to gather data relating to many other disciplines, from particle physics to new drug developments, both of which can have major impacts on society today or tomorrow — and where there may be no location attribute at all or where that attribute is only of small value or never changes. We contend that GI is not unique simply because many in the GI community (which also needs to be defined) decide that it is expensive to collect, use, or share.

Is GI unique because of the impact its use can have on society? This is perhaps the one area where GI has a claim to some degree of importance and uniqueness over many other forms of information, or at least the location attribute does. Accurate spatial attributes applying to numerous classes of information help to plan, operate, and maintain many other forms of societal infrastructure, either principally or subordinately, i.e., transport, food production, health, education, and many areas of governance generally, at all levels of government. The same cannot be said for multimedia content — other than perhaps for the degree of employment that the other information market sectors generate. A study conducted for the Ordnance Survey of Great Britain in 1999 reported that in 1996, Ordnance Survey (OS) products and services "contributed to 12–20% of gross value added (GVA)" amounting to "£79–£136 billion worth of gross value added (GVA)" mainly because of the use of "OS products and services as a primary input into production by several key sectors in the economy (e.g. utilities, local government and transport)" (OXERA, 1999). The key word in the above quote is the *to* in "contributed to." The report does not claim that OS data availability generated £79 to £136 billion worth of gross value added, but rather that existence and use of high-quality OS data made a significant contribution to the realization of these GVA figures by other sectors of the economy. Many other cost–benefit studies, some of which are reviewed in Chapter 6, support this general picture; i.e., that availability and use of good-quality GI can provide several times the benefit compared to cost through impact of such use on different sectors of the economy.

1.3   Valuing information

What do we mean by value of information, especially in regard to GI? The issue is so important in underpinning the free-or-fee debate on funding access to and use of public sector GI that we devote the whole of Chapter 2 to the topic. In this section, we simply introduce some of the aspects of information value that indicate why such effort is needed later. First, the same information can have different values when used in different ways by or for different people, at different times, in different formats, or when used for purposes other than that for which it was initially collected. Many GI industry professionals note that information itself is of no intrinsic value, but that value is tied directly to use and the nature of that use by the value it adds to the decision-making process (Longley et al., 2001, p. 376; Barr and Masser, 1996).

A single item of data may be used in many different ways, each use creating new information, usually when combined with other data, which are then collectively referenced or analyzed in unique ways, depending upon the application and the user's information intelligence requirements. The commercial, monetary value of a data product or service is only one of many types of value that can be assigned to information, yet this value is not appropriate or applicable in many circumstances. Of course, in the private sector of the information industry, the monetary value of data must be sufficient to recover development, production, sales, marketing, and dissemination costs, preferably with a return on those investments; otherwise, the product will soon disappear from the marketplace. Thus, 1 km of road centerline (location) data collected by the original data holder may be sold to a user for X.xx euro, representing the producer's cost recovery and return on investment (profit) targets and the buyer's willingness to pay, which are the main determinants of market price for any good or service.

Yet the real value of that road centerline will vary considerably depending upon the final uses to which it is put. First, there is the value to the first buyer, who perhaps incorporated this centerline data into a new product or service, thus becoming a value adder, selling that product or service on to other users at a new price, set again by cost recovery, return on investment (ROI), and willingness to pay parameters of this new marketplace. Users of that value-added product or service will view the value — to them — of that centerline quite differently. For example, the value to a driver for planning a trip from A to B by looking at a map (paper or on screen) is quite different than the value to the provider of a GPS-enabled, in-car navigation system (and its users) vs. the value to a highway maintenance team or utility company doing work on or near that road. The unquantifiable, intangible, and sometimes secondary or vicarious value of that centerline data, using the example of the GPS-enabled navigation system, differs significantly between your average car driver, who simply wants to get from A to B as easily as possible, and the road accident victim in the back of an ambulance who needs to get to the nearest hospital as quickly as possible to perhaps save his or her life.

To look at value from all these different viewpoints requires a return to value theory itself, a review of various information value chains that have been proposed in the information market, and, finally, some consideration of the many different ways that value can be increased — or decreased — due to collection issues (accuracy, timeliness, currency, etc.), technical issues (data formats, presentation formats, interoperability, etc.), and access issues. On the way, we provide a brief overview of the information market and the role that information infrastructure plays in developing and serving that marketplace, plus various studies that attempt to assign a value to geographic information within society or to the economy as a whole, either directly or indirectly.

Our main conclusion is that so many different types of information can be labeled as geographic (as we saw earlier in this chapter) that it is exceptionally difficult to assign a value to GI in general terms. This is the first hurdle, even before we enter into the more complex discussion surrounding different types of value. We know that monetary value based on GI data sales revenue is only a very small part of the tale. Such sales figures are also less than indicative of what value society attaches to GI, since they include commercial data sales by private industry and sales of GI by governments at local, regional, and national levels, plus by one government agency to another, raising the issue of multiple accounting of the sale and perceived value of GI if one were to accept sales figures as a valid surrogate for value.

Nevertheless, numerous cost–benefit studies for many sectors of industry, in different economies, create or use GI and report benefit–cost ratios of 2:1 up to 150:1 for using geographic information. No such study, report, or specific case identified during the research for this book has ever reported a single negative benefit–cost ratio for GI. Perhaps a leap of faith is required — and justified — for investment at the national level in GI whose value is otherwise so difficult to pin down. Most spatial data infrastructure (SDI) projects or initiatives undertaken in the past or contemplated today require that a cost–benefit analysis (CBA) be produced prior to committing to the levels of investment forecast for many such initiatives. Accepting a stated value for various types of GI is crucial in most formal CBA methodologies, yet a cost–benefit analysis is only as good as the assumptions that underpin the analysis methodology, and where benefits are concerned, much depends upon the value assigned to the GI at the heart of the initiative. Value is also one of the determinants of pricing and charging regimes, as discussed in Chapters 3 and 4, and also has an impact on the cost assumptions in a CBA — a vicious circle. The guiding principle for value, generally, should be that unused information has no value to anyone, so it is essential to establish the access regimes, exploitation principles, and infrastructures that maximize use, without compromising on quality and sustainable availability.

1.4 The debate on charging for public sector geographic information

As stated in the introduction, the ongoing debate on charging for public sector geographic information (PSGI) revolves around access and exploitation rights, often with little consideration of the true value of different forms of PSGI. To charge or not to charge for PSGI becomes a binary debate of good vs. bad. The authors feel that it is time to progress beyond these entrenched, secular (mainly GI-focused) polarities to examine processes and trends in the evolution of the information society and information markets, within which GI is simply one component.

No one questions the right of commercial firms to charge for the information they disseminate, even though many data products are derived wholly or in part from data originally gathered by or for the public sector. Sometimes exploitation rights are acquired for free, and at other times costs are imposed by the data owners. Unless these costs are unduly onerous or the data owner's position as provider is abused, charging for exploitation rights has not been proven to be detrimental to production and sales of commercial information products. A counterproof is usually offered, i.e., "The GI market is much larger in country A, where PSGI data is free, than in country B, where it is not, so free PSGI must be better than charging for exploitation." Yet robust markets have developed for GI-based information products and services in countries where exploitation rights are not free and many of the GI products and services in for-free countries are based on costly additional data collection and processing by the value adders. This claim was confirmed as recently as May 2007 in discussions with several major GI data providers in the U.S. at a GIS vendor's conference. These value adders all claimed that access to the U.S. federal data had some impact on GI product development, but that this advantage disappeared very rapidly as they took on the role — and often considerable cost — of updating and maintaining key data sets that are only available from federal sources based on very long update cycles, typically 10 years or more, and sometimes of uncertain quality.

Should all public sector information (PSI), geographic or otherwise, be available free of charge to citizens, or is it possible to charge for PSI and still have fair distribution? What level of resources should a public sector body commit to converting data required for legally-mandated purposes into information useful to and usable by the average citizen, and in funding that dissemination? Such value-adding and publishing tasks are typically the role of commercial organizations that have the relevant skills, experience, and access to capital and distribution networks, in return for which they expect a profit. Yet value adders need access to the basic data.

Current PSGI access debates seldom progress beyond entrenched positions based on ideology and emotion wherein access policies are riddled with contradictions. The European Union's PSI reuse directive of 2003 (EU, 2003) promoted policies for maximizing access to PSI, implying that charges for access and reuse of PSI should not exceed costs of reproduction and dissemination. With regard to GI, this policy is constrained by the difficulties of funding an unknown demand for PSGI whose collection is supported solely from direct taxation for some major GI-producing government agencies. The PSGI owner often does not set government policy on access to PSI, and the most recent PSI- and PSGI-related directives permit a wide range of policies to be implemented by governments. We must try to differentiate between the value of information and the goals of dissimilar PSI charging regimes, no easy task for GI, as we have already seen how difficult it is to even define the term, and alluded to how difficult it will be to assign specific values, as presented in Chapter 2. A market value for GI may be determined by different market places, but charging regimes depend on wider government information policy, national information cultures, and evolving e-government initiatives.

The U.S. is the home of freedom of information (FOI) for taxpayer-funded PSI created by federal agencies. The U.S. Geological Survey (USGS) recently admitted that much of its topographic data has not been updated for more than 50 years (Brown, 2002), due to lack of financial resources, provided solely from direct taxation. USGS has entered into a nationwide program — The National Map — to integrate more recent and more accurate GI collected by state and local governments into the national database because it does not have the federal funds to do this job itself. In the U.K., Ordnance Survey of Great Britain data are integrated, spatially- and temporally-detailed, highly structured for use in GIS and other application tools, and available to users shortly after the updates are entered into the National Digital Topographic Database — at the rate of over 50,000 updates per day. However, these data are not available for free, even to other government agencies, local or federal, let alone to third-party value-adding commercial organizations. Looking only at these two examples, we see freely available and exploitable data on one hand, but of questionable value due to quality, coverage, and currency issues, compared to data available only for a fee, but of high quality, full coverage, and updated daily. Which is best — or is there a single best scenario?

The polarized debate, i.e., good vs. bad, on direct charging for data arises from the competing goals that all information should be available to everyone in an "information commons" vs. capitalist arguments and business strategies based on paying for what you value and need. Charging for information is a complex issue, confused by near-zero dissemination costs allegedly offered by the Internet, whose early proponents espoused free availability of information, the cost to be recovered from advertising. Although this business model is yet to be proven successful in the long run, more than a decade after it appeared, data consumers avidly embrace the concept of apparently free information, happily unaware that someone is paying for this access, somewhere, even if not the immediate user. Yet information is truly valuable only when it becomes knowledge, and today knowledge is being embedded into machines and algorithms and delivered via the information products and services those machines offer to nonexpert users. Everyone can read a map, but not everyone can work out the optimal route to get from A to B, bypassing traffic congestion and road closures and using only nonmotorway roads.

Some debaters invoke human rights and universal information access principles to justify unrestricted availability of PSGI. Data can both empower and disenfranchise citizens, depending upon their particular circumstances in being able to make the best use of the information. Charging for PSGI is often demonized and made to seem undemocratic, forcing exclusions in society at a time when overcoming social and economic exclusion is a key policy goal, and forcing behavior that focuses on ability to pay, not on need to use. These tensions also emerge when something is commercially lucrative, yet has such potential public good that it should not be commodified, e.g., knowledge of the latest virus spreading across the Web. Should this be free because it potentially affects hundreds of millions of Internet users, or should those who detect and disinfect such viruses be paid for this valuable service? If they are not paid, for how much longer will such services continue to be offered, regardless of how valuable they are to society — and would any government agency be able or allowed to try to match the efficiency of such market-driven initiatives?

The conceptual basis of an anticharging argument is that it necessarily leads to a form of prejudice against those who cannot readily pay for access. In societal terms, this is linked to policy interventions regarding social exclusion, often expressed as the problem of overcoming the digital divide. This wider debate encompasses the fundamental question of whether information and data, above anything else, should be made freely available.

The authors are not yet persuaded that information alone can have a direct impact on public good, which is why we feel that the debate is important — and unfinished. In a digital society, many people assume that information reproduction and dissemination costs are nearly zero, so the data should be freely available. Yet producing socially and economically useful information incurs real reproduction costs involving infrastructure, machinery, and skills. Policy interventions attempting to correct the digital divide show how difficult it is to decide where benefits from free availability end and benefits from charging begin. A subsidy may be needed that invests in developing information literacy that empowers people to make sensible use of the information. Governments could interpret this subsidy not as a process of leveling market distortions, but as an investment in social and intellectual capital formation.

Developing nations face different problems, where governance reform is required. In many such states, PSGI is either nonexistent or sadly out of date, requiring significant investment to improve the situation — investment not available from current government resources. We need to extend the debate to the development arena, to explore how it can articulate the challenges and tensions facing countries with low levels of existing PSGI productivity and limited government resources, and attempt to create spatial data infrastructures (SDIs) that will help them compete globally. What is the answer to the let the government/taxpayer pay once solution when there is no money in the treasury to make those payments?

Pricing policy depends upon complex relationships between users and suppliers, between the perceived value of primary PSGI and possible substitutes. Better understanding of the value of GI may alter the pricing policy debate, but we have already seen the confusion that exists over such basics as the definition of the term *geographic information*, let alone the different value issues. Confusion will increase as governments outsource more of their data collection, processing, and dissemination workloads — activities for which policies already vary among governments. Tensions will remain concerning the extent to which PSGI producers can generate sufficient capacity from selling data, services, and value-added products to satisfy demand, especially if new demand arises as a result of making PSI more widely available and more fully exploitable. Tensions will remain surrounding fears of unfair competition and monopolistic control over the supply chain that arise from near-monopolistic supply of PSGI by a single, legally-mandated government agency.

We need to reorient the charging debate away from entrenched dogma to look more objectively at charging regimes based on economic reality and true value to all members of society, reducing some of the near-religious fervor attached to concepts of information freedom, civil rights, public goods, and information commons. Rearticulating the charging debate as one of the differential strategies to build capacity in an uncertain environment focuses on doctrines that best achieve flexibility and quality of user service. Some countries or regions have already experimented with charging vs. no-charging regimes — and even switched back and forth over a number of years. We have much to learn from these real-world experiments, i.e., what worked and what did not — and why not? The time for gut reactions that free must be good is over; it is now time to look at what is actually happening in the real world, without prejudicing the outcome of such debate.

1.5 Overview of the contents

Chapter 1 has given you a taste of the fee-or-free debate and introduced the terminology used. This section presents an overview of the remaining chapters, with a very brief summary of the main conclusions from each chapter. From this description, we hope that readers can dip into those sections of the book of most interest to them — or to their part in the fee-or-free debate.

Chapter 2, "Determining the Value of Geographic Information," introduces a range of issues impacting on value of information generally and the value of GI specifically. The chapter explores different measures of information value, value theory applied to information, different types of information value chains, the information market, different components of value inherent in GI, and the value of GI to economies and society. We explore to what extent the location attribute of GI is the fundamental hook on which other data, such as official statistics or health information, can be structured, analyzed, and used. Just how valuable is the location attribute for otherwise nontopographic (land features) information? Can a meaningful value be placed on the location attribute alone for a specific piece of information in isolation to the value of other valuable attributes of that information, even for topographic data? The example of information on crop growth rates is introduced to portray the many sides to the value question, such as how valuable, to whom, when, for what purpose, and in what format? What happens in the value chain when private sector GI becomes public, i.e., when governments outsource data collection, and vice versa, i.e., when private firms exploit public sector GI?

The chapter concludes that there are many ways to define value, some quantifiable monetarily and some not, even for the same piece of GI or PSGI, depending upon the reason for which it was initially collected or needed and the eventual circumstances in which it was used. Thus, value — or rather perceived value — may not be a meaningful guide to pricing or charging for PSGI where this practice is followed by some governments, or as an automatic justification that all PSGI should be made freely available in defense of the information commons argument. This inability to assign consistent and persistent value to various types of GI has a direct impact on the cost–benefit strategies adopted to justify SDI investments, as discussed in Chapter 6.

Chapter 3, "The Business of GI: No Such Thing as a Free Lunch," focuses on the interplay of price, cost, and value, introducing the reader to the main elements in the pricing and charging debate surrounding public sector GI (PSGI). Key questions include:

- Should PSGI be made freely available as part of the information commons?
- Can PSGI be made available for free, and what is the potential impact of that decision on the quality of the PSGI itself and on the public purse?
- If PSGI is made freely available — and for free — can this informational "free lunch" be maintained into the future without negatively impacting on information quality?

The chapter explores different and evolving business models for making information freely available, from online publishing of freely available peer-reviewed journals to Wikipedia to hybrid models of "some data free, some for a fee," and whether these are applicable to PSGI provision. The impact on quality of topographic PSGI in the U.S. under the "free to all, for any use" policy of U.S. federal government agencies is examined. The debate on charging for PSGI is dissected into political myths and funding strategies required to build information production and maintenance capacity in uncertain funding environments. The continually changing power relationships in the PSI rights of access debates are explored, noting that the free lunch arguments in these debates are made mainly by those who have the most to gain from free access to data, not necessarily always to the benefit of the data-providing organizations.

The chapter concludes that the value of PSGI to the information market is not only related to the cost–benefit of using GI in myriad applications, but also in the potential investment in GI as a market in its own right. Most government GI producers are not independent operators, and their activities are deeply constrained by government policies. These policies are subject to sudden and unexpected changes, just as the economy is subject to changes through the processes of globalization. Providing access to PSGI is an economic and political contest between resource allocation, constrained by finite budgets, and competing user demands.

Chapter 4, "Pricing Information: The Interaction of Mechanism and Policy," explores and extends pricing theory as it relates to information and examines several issues surrounding pricing of PSGI and the changing relationships between information producers and consumers. The chapter begins with a review of first-, second-, and third-degree price discrimination, then proposes an extension to zero-degree price discrimination to accommodate the free lunch debate of Chapter 3. Practical examples of the consequences of underfunding for national mapping agencies are presented as a warning to how PSGI quality can suffer under the free lunch approach to PSI provision. Various GI market positions are examined, including first-mover advantage, legacy systems, and monopoly supplier issues. Pricing contexts and strategies are discussed in relation to different types of costing scenarios, e.g., subsidy, contribution, absorption, or indirect costing.

The chapter concludes with the observations that the tendency for some public sector GI data producers to charge for their data is neither special nor new in the context of the information content industry today. Even those national mapping agencies who strive to make most of their GI available for free or at minimal cost, to most users, typically then also offer value-added products or services of their own or restrict use of the freely available data for commercial purposes. Laudable examples of attempts to build the GI information commons exist, such as the Open Street Map project in the U.K. and similar citizen-based mapping projects around the globe, yet such initiatives are unlikely to seriously threaten established players, unless they reach a critical mass, such as Wikipedia did in 2006. The wide range of examples introduced in the chapter indicates that economic, political, and social turbulence in the pricing of information is growing, not decreasing. PSGI free lunch experiments exist and will continue, but face difficulties in regard to centralized funding by government, no matter what the emotional and economic arguments are about justification and need.

Chapter 5, "Geographic Information, Globalization, and Society," explores the nature and role of GI in today's society, beginning with the myth that GI is ubiquitous, a fundamental component of all information, then moving on to the politics of information, development of spatial data infrastructures (SDI), and privacy and surveillance issues. The globalization of GI is examined in the contexts of mobility, location, and boundaries. We propose that the repurposing of GI is affected by the acceleration of processes across space, and by an increasing sophistication of repurposing use of GI. This has extended the GI supply chain beyond owning and using data, to a more sophisticated and demanding dependent producer–user relationship in which it is increasingly difficult for GI producers to understand the extent of the repurposing of their data. New and diverse user demands exist because of the sophistication of the GI market, demands that go well beyond simply reformatting once familiar content into new media forms, to the production of new types of data and applications.

Yet another paradox that emerges through the wider availability of GI is the extent to which the volume of information is creating "noisescapes" through which citizens have difficulty navigating. This is due not only to information overload, but also to the complexity of debates to which citizens are exposed. The provision of increasing levels of environmental information also introduces complex feedback effects, which are to a large extent circular. With ever more information accessible to citizens, even more information about those citizens and their localities is available to business and government.

The chapter investigates the impact of technology and intellectual property rights (IPR) on GI consumption in society. Using GI requires tools and techniques that together comprise the technologies of geographic information systems (GISs). In the twenty-first century, just as information is becoming more readily accessible, many familiar and common knowledge techniques are becoming less accessible through the privatization of knowledge via the patent system. Because theft of information for IPR is easier in the digital age than for print media, IPR protection laws have become more restrictive, with often unexpected or unwanted side effects on society's ability to use the new digital information resources and services. This trend is evident not only for GI, but also for music, books, and other forms of media, and has spawned its own series of bipolar debates on what actions should be permitted in regard to use of a digital media product once acquired by a consumer. Existing approaches to IPR protection are uneven and will continue to impact significantly on the availability of GI and the tools and techniques required to process GI most effectively. From the societal point of view, the conclusion is that we are in a time of divergent trends, increasing GI production, some selective censoring of GI, and an increasing monopolization of many essential techniques that people need to use GI. Thus, the twenty-first century promises to be just as turbulent as was the latter part of the twentieth century, and separate debates will continue relating to IPR issues other than simply fee-or-free access to GI.

Chapter 6, "Spatial Data Infrastructures: Policy, Value, and Cost–Benefit," extends the discussion on spatial data infrastructures (SDIs) begun in Chapter 5, looking more closely at the policy issues and strategies being adopted to create SDIs generally; recognizing that policies set goals and principles and strategies form the basis for implementation of policies. Special concerns relating to implementation of SDI policies are discussed, for example, relating to the supposed ubiquity of GI within the government PSI sphere and the added complexity this brings to the governance of SDI. Because GI spans numerous — perhaps all — government sectors, identifying a single lead agency to be in charge has proven difficult at both national and transnational levels, often delaying SDI implementation due to lack of ownership of the initiative. Examples of current SDI policies globally are presented and dissected, including the role of information access and pricing principles and policies within existing SDI initiatives at national and regional (transnational) levels, the latter focusing on the Infrastructure for Spatial Information in Europe (INSPIRE). Finally, various cost–benefit issues are explored in relation to justifying SDI implementation budgets, tying back to the discussion in Chapter 2 on the value of GI and the difficulties that assigning value to GI creates in regard to the cost–benefit equation for information infrastructures. The chapter concludes with some observations on which cost–benefit methodologies might be most effective in justifying to senior decision makers the potentially high investments for SDI development.

Chapter 7, "Conclusions and Prospects," opens with the authors' conviction that the current and expected continued fluidity of the information landscape prevents drawing any firm conclusions at this stage, when existing beliefs and practices regarding information markets and infrastructure are being challenged regularly and continue to evolve. Instead, we provide some observations and conceptual summaries that may help explain where we have come from and why, and hopefully offer some insight into where we are going in the future. We do not propose that there is one simple answer to the fee-or-free debate in regard to access to, and exploitation of, public sector

GI (PSGI). Rather, the case for free information can be made on the basis of freedom of information principles or for the public good and delivering public value, similar to recent calls for wider free access to publicly-funded research through an information commons.

Some governments have made the strategic decision to release PSGI freely for the wider public good; for example, Canada and the government of Catalunya, Spain, at the national and subnational levels, respectively. However, success of such initiatives will depend upon maintaining a sustainable funding stream, dependent upon the goodwill of the government of the day to provide the funds needed by data providers, and the ability to match data provision to market needs. A 2001 study in the Netherlands (Berenschot & NEI, 2001) on financing public sector databases proposed that the middle ground was best, i.e., prosperity effects would be maximized when public sector data were sold at marginal cost, not given away freely. But as Mike Clark commented in 2006, though NEI-Berenschot stated that lower prices could possibly lead to more users, "increased use does not automatically mean either increased tax revenue or decreased societal cost" (Clark, 2007). NEI-Berenschot acknowledged that it was "actually impossible to reasonably quantify economic prosperity effects at the macro level," and that reducing existing charging levels would always, in the short term, increase the burden on the already typically overstretched public purse (Clark, 2006).

We observe that the current fee-or-free contest is not unique to the GI sector, but appears in regard to public provision of transport, health, and communications infrastructure, where there is increasing evidence in many countries of a move away from provision via subsidy to pay-for-use, especially when the subsidy proves to be inadequate to meet demand arising from the free-access regime. For us, the key point is the availability of consistent resources for reinvestment and maintenance of information that is fit for a wide range of purposes, which simultaneously maximizes the ability of information providers to respond to the widest possible user base and market.

We close the book with the following conclusions and observations, which seem to apply reasonably well on a global scale:

- There is a growing mismatch between organization speed and market speed.
- The importance and role that public sector information (PSI) plays in the economy will continue to be strong through its role in allocation of government resources and the measurement of government performance.
- National-level PSI will continue to be contested concerning its relevance and quality in relation to local level needs.
- The threat will continue to grow whereby PSI is collected by government, directly or by subcontract, but where the only users of the data are organizations that are mandated to use the data through an official "process monopoly." This potentially dangerous situation needs to be monitored and, if necessary, perhaps more closely regulated.

- Challenges to the information and knowledge commons through the uncertain exercising of monopoly patents on a global scale will continue.
- The process of making geographic information available will engender ever more flexible strategies in the future.

In best textbook fashion, we leave it to the reader to put these observations to best use, comparing them to the situations they find in their own organizations, regions, sectors, or nations. Most importantly, we ask that those reading the following chapters do so with open and critical minds. It is time for the debates on how to best fund the collection and use of public sector geographic information (as we defined GI earlier in this chapter) to be raised a level, to include more open, evidence-based, and objective rationales.

References

AGI. 1991. *GIS Dictionary: A Standards Committee Publication*, Version 1.1. The Association for Geographic Information, London. http://www.geo.ed.ac.uk/agidexe/term?1735 (accessed March 20, 2007).

AGI. 1999. *GIS Dictionary* (online). The Association for Geographic Information, London. http://www.geo.ed.ac.uk/agidexe/term?43 (accessed January 15, 2007).

ANZLIC. 2006. Glossary of Spatial Information Related Terms. ANZLIC Spatial Information Council website. http://www.anzlic.org.au/glossary_terms.html (accessed May 20, 2005).

Baltimore County (Maryland). 2001, November. *Cost Benefit Geographic Information Systems*. Baltimore County Office for Information Technology Business Application Unit Geographic Information Systems.

Barr, R. and I. Masser. 1996. The economic nature of geographic information: a theoretical perspective. In *Proceedings of the GIS Research UK 1996 Conference*, GIS RUK '96, University of Kent, April, 1996, 59–66.

Berenschot and Nederlands Economisch Instituut, 2001. Welvaartseffecten van verschillende financier ingsmethoden van elektronische gegevensbestanden. Report for the Minister for Urban Policy and Integration of Ethnic Minorities, Government of the Netherlands.

Blinn, C., L.P. Queen, and L.W. Maki. 2007. *Geographic Information Systems: A Glossary*. University of Minnesota Extension. http://www.extension.umn.edu/distribution/naturalresources/components/DD6097ag.html (accessed April 10, 2007).

Booz Allen Hamilton. 2005. *Geospatial Interoperability Return on Investment Study*. National Aeronautics and Space Administration, Geospatial Interoperability Office. www.isotc211.org/Outreach/Newsletter/Newsletter_08_2005/TC_211_Newsletter_08_Appendix_I.pdf (accessed September 8, 2002).

Brown, K. 2002. Mapping the future. *Science*, 298: 1874–1985.

CEN. 1998, November. *Geographic Information: Vocabulary*, CR 13436:1998. CEN Secretariat, AFNOR, Paris.

CIE. 2000, March. *Scoping the Business Case for SDI Development*. Center for International Economics, Canberra and Sydney, Australia.

Clark, M. 2007. Fee or Free? The hidden costs of free public sector information. Business Information Review, 3/2007, vol. 24 pp. 49–59.

Craglia, M. and INSPIRE FDS Working Group (Eds.). 2003, September 24. *Contribution to the Extended Impact Assessment of INSPIRE.* Environment Agency, England and Wales.

DOE, U.K. 1987. *Handling Geographic Information.* Department of Environment, Her Majesty's Stationery Office, London.

ESRI. 2001. *Dictionary of GIS Terminology.* The ESRI Press, Redlands, CA.

ESRI. 2007. *GIS Dictionary* (online). http://support.esri.com/index.cfm?fa=knowledge base.gisDictionary.gateway (accessed March 15, 2007).

EU. 2003, December 13. Directive 2003/98/EC of the European Parliament and of the Council of 17 November 2003 on the Re-Use of Public Sector Information. *Official Journal of the European Union* (Luxembourg). http://www.psi.gov.ie/psi-documentation-background-information/19052005-eu-directive-final-text. pdf/ (accessed March 10, 2007).

EU. 2007, April 25. Directive 2007/2/EC of the European Parliament and of the Council of 14 March 2007 Establishing an Infrastructure for Spatial Information in the European Community (INSPIRE). *Official Journal of the European Union*, Luxembourg. http://www.ec-gis.org/inspire/directive/l_10820070425en00010014. pdf (accessed April 30, 2007).

FGDC. 2007. *Content Standard for Digital Geospatial Metadata*, Appendix A, Glossary, FGDC-STD-001-1998. Federal Geographic Data Committee Secretariat, Reston, VA. http://www.fgdc.gov/metadata/csdgm/glossary.html (accessed January 21, 2007).

Garribba, M. 1999. European content on the global networks. In *Proceedings of Electronic Publishing 1999 Conference, Redefining the Information Chain: New Ways and Voices*, Ronneby, Sweden, May 10–12, 1999. http://www.bth.se/elpub99/ap.nsf/08c6c2f88424ad99c12566ff002a0c10/5d3286fba8905dc0c12567980035d912/$FILE/287-292.pdf (accessed April 30, 2007).

GIS Development. 2007. The Geospatial Resource Portal. http://www.gisdevelopment.net/glossary/g.htm (accessed February 10, 2007).

Halsing, D., Theisen, K., and Bernknopf, R. 2004. *A Cost-Benefit Analysis of The National Map*, Circular 1271. U.S. Department of the Interior, U.S. Geological Survey, Reston, VA. http://nationalmap.gov/nmnews.html.

Hardwick, P. and B. Fox. 1999, February 10. *Study of Potential Benefits of GIS for Large Fire Incident Management.* USDA Forest Service by Pacific Meridian Resources.

ISO. 2002. *Geographic Information — Reference Model*, ISO 19101:2002. International Standards Organization, ISO/TC211 Geographic Information/Geomatics Secretariat, Norwegian Technology Centre, Oslo, Norway. http://www.iso.org/iso/en/CatalogueDetailPage.CatalogueDetail?CSNUMBER=26002 (accessed February 20, 2007).

Longhorn, R. and M. Blakemore. 2004. Re-visiting the valuing and pricing of digital geographic information. *Journal of Digital Information*, 4: 1–27. http://jodi.tamu. edu/Articles/v04/i02/Longhorn/ (accessed March 7, 2007).

Longley, P., M. Goodchild, D. Maguire, and D. Rhind. 2001. *Geographic Information Systems and Science.* Wiley & Sons, Chichester, U.K.

Maguire, D., M. Goodchild, and D. Rhind. 1991. *Geographical Information Systems: Principles and Applications.* Longman Scientific and Technical, Harlow, U.K.

Montgomery County (Maryland) Council. 1999, April 12. *Geographic Information System Cost/Benefit Assessment Report.* Montgomery County Council's Management and Fiscal Policy Committee.

Oxford Economic Research Associates Ltd. (OXERA). 1999, May 14. *The Economic Contribution of Ordnance Survey GB*. For OSGB by author. http://www.ordnance survey.co.uk/aboutus/reports/oxera/oxera.pdf (accessed September 8, 2007).

PIRA. 2000, September 20. *Commercial Exploitation of Europe's Public Sector Information*, executive summary and final report. PIRA International Ltd. and University of East Anglia for the European Commission, DG Information Society. http://ec.europa.eu/information_society/policy/psi/docs/pdfs/pira_study/2000_1558_en.pdf (accessed Sept. 8, 2007).

Price Waterhouse. 1995. *Australian Land and Geographic Data Infrastructure Benefits Study*. For Australia New Zealand Land Information Council (ANZLIC) by author. http://www.anzlic.org.au/get/2358011751.pdf (accessed Sept. 8, 2007).

Prodger, D. and N. Sutherland (Eds.). 1997, February 11. *GI-Base*, Draft Final Report, Ordnance Survey GB for DG XIII of the European Commission.

Waltham, M. 2002, March 7. As the Future Unfolds, What about the Past? SPI Seminar. http://www.marywaltham.com/Archiving_talking_notes.htm (accessed May 9, 2007).

Werschler, T. and Rancourt, J. 2005, Winter. *The Dissemination of Government Geographic Data in Canada: Guide to Best Practices*, Version 1.2. GeoConnections Canada, Ottawa, Canada. http://www.geoconnections.org/publications/Best_practices_guide/Guide_to_Best_Practices_v12_finale_e.pdf (accessed Sept. 8, 2007).

chapter two

Determining the value of geographic information

2.1   Introduction

Everyone is a user of information, and the same information can be used by all sections of society for quite different purposes — citizens, businesses, and public bodies. In this chapter, we address the question: What is the value of geographic information? Longley et al. (2001, p. 376) note that "the value of the same information differs hugely to different people and for different applications." Different values also apply at different times or when information is in different formats or when used for purposes other than that for which it was first collected. According to Barr and Masser (1996), "information has no inherent value, it is only of value once used and that value is related to the nature of the use rather than the nature of the information. As a result information has very different values for different users." According to the U.S. Federal Highway Administration (1998, p. 3), information has value "determined by its importance to the decision maker or to the outcome of the decision being made ... professionals require information that is not only accurate, timely, and relevant, but also presented and interpreted in a meaningful way." To complicate matters, as we saw in Chapter 1 (p. 2), geographic information (GI) has many definitions.

The very meaning of the word *value*, in relation to worth, is another indication that it may be extremely difficult, if not impossible, to assign any one value to something as multifunctional and multifaceted as information.

The Many Meanings of *Value*

Value, noun, worth; intrinsic worth or goodness; recognition of such worth; that which renders something useful or estimable; relative worth; high worth; price; the exact amount of a variable quantity in a particular case. (Larousse, 1997)

Value, noun, the importance or worth of something for or to someone; how useful or important something is; the amount of money that can be received for something. (*Cambridge Advanced Learner's Dictionary*, 2005)

Value, noun, a fair return or equivalent in goods, services, or money for something exchanged; the monetary worth of

> something, e.g. a market price; relative worth, utility, or impor-
> tance; a numerical quantity assigned to something or determined
> by calculation or measurement. (Merriam-Webster Dictionary
> Online, 2007)

Value of information or information-based services seldom relates to purchase price or cost, except for the monetary value received by a vendor from
sale of information or services. However, the value perceived by a customer
may impact on the price charged by a vendor or the customer's willingness
to pay. In the commercial marketplace, for an information product or service
to be sustainable, price must cover at least cost of production and distribution, and preferably some return on investment.

For public sector geographic information (PSGI), required or produced as
part of a public body's governance responsibilities, any value based on commercial price to acquire data or a service may be irrelevant, since the data
must be collected or used in order to fulfill legally-mandated tasks. In this
case, the true value to both the public body and society, i.e., citizens and businesses, lies in the efficient completion of those tasks. For both the public and
commercial sectors, remember that all information has a cost, yet the cost
for acquiring and using the same information may vary, and the same information may have differing values for different users at different times, in
different formats, with different conditions attached. As Bryson (2001) notes,
it becomes important in the global information society to "identify and manage different value propositions from a financial, political, corporate, social,
cultural, personal and community values perspective ... to exploit the total
worth of the information and knowledge age." Also, Lash (2002) introduces
the concepts of exchange value and use value, in which use value typically
exceeds exchange value.

A warning is perhaps in order here for the reader who is looking for in-
depth coverage of the many issues surrounding value of information. This
chapter provides an overview of the issues and theories surrounding the
definition of value, many of which warrant entire books in their own right
— and indeed some of the topics, such as value theory, value chains, and
information economics, have generated entire literatures. Therefore, we have
limited ourselves to setting out the key issues and definitions, and introducing the reader to some of the underpinning theories, which can be explored
more fully using the extensive references listed at the end of the chapter.

2.1.1   *Information value is in the eye of the beholder*

The value of information as a product, sold by a vendor, may not equate to
the value of that same information to the final consumer or user. For the
former, the value of information may be totally financial, based on a sales
price that covers all costs plus an acceptable return on investment. For the
user, depending upon the type of user, the value might be financial, social,

economic, cultural, political, or personal, as Bryson (2001) indicated. At the personal level, the value could vary from simple added convenience, e.g., finding a restaurant or theatre more easily, to enabling a new information service offered by the user for his or her financial gain. Also, what is the value to a vicarious user, i.e., the value of location-based data used in an emergency vehicle routing system that may help save a person's life — your life? Thus, one can see that the question "What is the value of GI?" depends very much on who is asking and why. A GI vendor who is making an acceptable profit from sales of a GI product or service is quite happy with the value of the GI on offer. A purchaser disappointed by the utility that he or she received from that product or service, for a specific purpose in certain circumstances, might be less inclined to assign high value to the very same GI.

Disregard for the moment the distinctions typically made among data, information, and actionable knowledge gained from use of information. Set aside the claim that "geospatial information is special" (Van Loenen, 2006, p. 19) in the world of information and information markets. As mentioned in Chapter 1, some of the aspects of GI put forward to support the claim for the uniqueness of GI also apply to many other types of data, especially in the scientific, technical, and medical (STM) realm. Regarding perceived value, this ephemeral thing called information has similarities to physical goods that one can see and touch. For example, a chair has production costs, which must be met by someone, as does information. A chair is created with some purpose or planned use in mind, some marketplace, as are information products and services. The chair may have different values to different people, e.g., a chair constructed in a 1950s' style might be desired by certain collectors of furniture from that period, and thus of high value, but considered to be hopelessly old fashioned by others, and thus of low value. The monetary value placed today on a Louis XVI antique chair certainly bears no relationship to its production cost. Similarly, geographic information describing road centerlines is of critical importance for a highway authority, and therefore of great value, but of little importance to a forestry commission, and of no use to a mariner, for which it is unlikely to have any value at all. Yet all three — highway authority, forestry commission, and mariner — are users of geographic information. Thus, while the value proposition may be similar between information and hard goods, the economics of information are quite different from those of physical goods, since "information can be costly to produce, but cheap to reproduce" (Longley et al., 2001, p. 379), and even less costly to distribute, especially in the digital age.

2.1.2 *What type of value to measure?*

Value should be measurable in some acceptable way. However, if information has different types of value, representing different aspects of worth, then there will be different measures, which will not apply equally to all information in all circumstances. One measure of worth is financial or monetary value, i.e., sales value related to production cost recovery, profit margins, and return on investment or similar financial targets within the commercial information market environment. This relates more to what Lash (2002) refers to as exchange value. Financial value can also apply to public sector GI if use of the information helps deliver cost savings or aids in managing financial risk while improving service delivery. In this case, however, the numeric value may be more difficult to specify and no longer necessarily relates to exchange value. Monetary value recognizes that information production costs are real, e.g., for data collection, processing, dissemination, and management, and must be recovered by someone, somehow. This type of value applies to raw data, as a commodity to be traded, and to value-added information products and services. Since costs can usually be computed with some degree of accuracy, this type of value, typically reflected in the price at which the data are traded and the consumer's willingness to pay for the product or service offered, can also be determined reasonably well. In other words, the sales price offered in the information marketplace serves as a financial surrogate for one type of value. Remember that both raw data and value-added products and services can have different perceived values to consumers, represented by the customer's willingness to pay. If this value is lower than production costs, then the data, product, or service will soon disappear from the marketplace.

Much is also written about the socioeconomic value of information, i.e., value of an information good or service in achieving societal goals, typically by impact on quality of life or better governance or improved economics at the macro level. Socioeconomic value is much more difficult to quantify than monetary value because of the myriad uses to which the same information product or service can be put in regard to a wide range of societal goals or economic targets. In this chapter, we review some past attempts to assign socioeconomic value to geographic information, for which the location attribute supposedly adds specific value. However, such value assignments are often frustrated by difficulty in translating acceptable measures of success in achieving often intangible benefits to society as a whole into something quantifiable, such as a monetary value or other tangible benefit for which a surrogate monetary value can be assigned. Proponents of GI as a valuable information resource often rely on such financially indefinable or ambiguous benefits when promoting the concept and value of spatial data infrastructures (SDIs) to government, for which costs at the national level can be considerable, an issue explored further in Chapter 6.

There is also the question of whether one should assess social value and economic value separately. According to Angeletos and Pavan (2007), research into the social value of information goes back more than 35 years, with the early work of Hirschleifer (1971), during which period competing claims are offered that "public information can reduce welfare (and) ... public information is necessarily welfare improving." In their 2007 (p. 568) paper, they show that "the social value of information depends not only on the form of strategic interaction, but also on other external effects that determine the gap between equilibrium and efficient use of information" (Angeletos and Pavan, 2007, p. 5). Their work investigates economies in which welfare (a measure of social value) would be greater if agents (decision makers) increased their reliance on public information, contrasted with economies in which just the opposite is true. They also describe economies in which any and all information is socially valuable contrasted with economies in which welfare decreases with increased access to both public and private information, the latter claim calling into question claims of the importance of the information commons to society.

Information also has cultural value, which may be considered separately from social or economic value, yet this is difficult to measure except in social terms, for which, as already indicated, it is inherently difficult to assign a specific value. Thus, cultural value is perhaps the most difficult of all types of worth to assign to GI or, for that matter, to other types of information and a whole range of physical objects, from historic monuments to the *Domesday Book*. Yet when one looks at the often significant sums that nations assign to cultural budget lines, e.g., for museums, libraries, orchestras, or maintenance of national monuments, it appears that culture is considered to be a valuable national asset. Information both protects and promulgates cultural identity, where place is a key attribute for much of the information deemed to be cultural. Information defines cultures, imparting a sense of identity, sovereignty, principles, and rights to those in a specific society, and also separates subcultures. One aspect of cultural value for GI relates to preservation of information, for example, of old maps or other place-based collections of data, which help us to understand human history and our place in that history, in our own society and in the global society, both today and in the past.

Defining what constitutes cultural information and the cultural values that relate to measures of worth, importance, or usefulness is no simple task, as cultural value is very closely linked to the social value of information and its supporting technologies. Again, according to Bryson (2001, p. 5), this is "because information and its supporting technologies assist with developing individual and collective minds and manners, and contribute to the intellectual and artistic development of different societies and groups." Understanding the rights of others is also one of the cultural values quoted by Bryson, which includes the right to determine "ownership, presentation and management of information and knowledge." In fact, much of our cultural heritage is captured in, or represented by, artifacts from our past, of all shapes and forms, including the information needed to interpret those artifacts in a cultural or societal setting. In that sense, geographic information provides cultural contexts, whether represented by the earliest maps, which were often produced as works of fine art, or simply textual references to events, objects, and people that establish spatial references.

Bryson also proposes that the political value of information derives from its usefulness in communicating ideas, principles, and commitments. We are all aware that information is used — and sometimes misused — by individuals, political parties, or nongovernmental organizations to promote specific viewpoints, usually to sway our opinions — or votes — one way or another over often contentious issues. For example, GI, or rather, the location attribute of much information used in urban and rural planning, is often key to various conservation organizations for achieving their aims for land or heritage preservation, often aligned against powerful and well-funded commercial property developers. Where the decisions made, or the issues discussed, have an obvious spatial context, such as locating a new housing development in the middle of a site of special scientific interest, the GI takes on a separate political value in its own right. If the spatial relationship attributes are used effectively, the political value of GI can be a powerful persuader. Sadly, as with much information, GI can be used for ill as well as for good, and such potential misuse then diminishes its political value not only in the instance where such use is detected by decision makers, including ordinary citizens, but perhaps in future similar situations as well.

Political value of GI can also be seen in the way its use can influence the interests, status, or even economic viability of organizations and individuals, when it is used to manipulate a specific outcome or to promote a particular viewpoint, or indeed simply to provide place-based information that can have both positive and negative impacts. For example, the high-resolution digital terrain model produced by one U.K. insurance company to be better able to assess flood risk nationally was of high positive value to the company and its shareholders, but of negative value to those former or potential policy holders now refused flood protection insurance if their property was located in a geographic area determined by the new model to be at high risk of flooding. At the same time, the availability of that new data set, whether made freely available or at an affordable cost, provided an important new GI resource for numerous governmental and private organizations involved in flood planning, remediation, and disaster management, certainly an added positive value for society.

Public goods are defined as any good that is nonrivalrous, i.e., "consumption of the good by one individual does not reduce the amount of the good available for consumption by others" (Wikipedia), and information is often used as a classic example. The term is also used to refer to goods that are nonexcludable, i.e., individuals cannot be excluded from consumption of the goods, although goods that are both nonexcludable and nonrivalrous are also sometimes called pure public goods. The economist Paul Samuelson is credited with developing the theory of public goods, defining a "collective consumption good" in a 1954 paper on the theory of public expenditure, as "[goods] which all enjoy in common in the sense that each individual's consumption of such a good leads to no subtractions from any other individual's consumption of that good" (Samuelson, 1954, p. 387). Many proponents of free access to GI collected by government, or indeed to any public sector information, base their belief on the principle of such information constituting a valuable public good, to be shared with all citizens on equal terms. Yet some economists also argue that total reliance on public goods can lead to market failures when such goods cannot be provided in sufficient quantity to satisfy demand. Tyler Cowen (2002, p. 1) proposes that "imperfections of market solutions to public goods problems must be weighed against the imperfections of government solutions. Governments rely on bureaucracy and have weak incentives to serve consumers. Therefore, they produce inefficiently." Onsrud warns against trying to set a commodity type value to data, information, and knowledge that are necessary for communicating at all levels and supporting democratic processes. He claims, rather, that information possesses the "classic characteristics of 'public goods'" (Onsrud, 2004). Weiss concluded that "public good characteristics" are one of the "fundamental economic characteristics of information" along with high elasticity of demand (Weiss, 2002). The role of the public good value in relation to pricing and charging for public sector information (PSI) and public sector geographic information (PSGI) is explored more fully in Chapters 3 and 4.

2.2   *Valuing Geographic Information*

Consider that the term *geographic information* has numerous definitions and manifestations, as described in Chapter 1. Satellite imagery of the whole earth, or even Mars, is geographic information that drives a multi-billion-dollar global satellite construction and space imaging industry. The virtual representation of real-world features such as the location of the centerline of a road or the bounds of a meandering riverbank, portrayed in some visual way in relation to other features, using a known coordinate system, is geographic information. The official (legal) boundary line of your property as recorded in a land registry database, which may or may not match the actual on-the-ground fence line separating your property from your neighbor's, is geographic information, just as is the location of that actual fence line. Such discrepancies between real-world and manufactured boundary data can have important legal, economic, and even political impacts, for example, where the discrepancy involves a national border. Man-made administrative boundaries, such as electoral wards, census enumeration districts, offshore economic zones, or boundaries created by marketing organizations for collecting and analyzing geodemographic data, all constitute geographic information, typically underpinned by artificial grid or coordinate systems. These boundaries establish the spatial referencing framework within which all the other attributes for the information of interest can be analyzed, whether it is household income, voting preferences, or the value of offshore mineral deposits. Finally, there are data describing objects or events using many attributes other than just location, for which the location attribute has different values depending upon who is using the data, how, when, and for what reason.

2.2.1   *Value changes with time, purpose, and use*

An image from space can have high value today, for example, in spotting the initial outbreak of a forest fire so that firefighting resources can be best allocated to save human life, property, and the environment. That same image will be of much less value tomorrow, or next week, once the fire has been extinguished. Yet the same image could regain value one year from now, or a decade or many decades in the future, as invaluable source material for analyzing environmental problems and trends. These include potential remediation (replanting) costs for deforested areas, the impact of deforestation on wildlife conservation and biodiversity, the potential impact on global climate change due to lost carbon sequestration capacity represented by the amount of forest destroyed. If existence of, and rapid access to, that initial image had resulted in a small firefighting team extinguishing a new fire in a matter of hours without significant loss of property, forest, or life vs. extensive losses that might occur without such advance warning, then what is the value of such information?

Further reflect on the changing value of information generated by repurposing of use. Imagery that underpins Google Earth™ or Microsoft's Virtual Earth™ online geospatial visualization services has acquired new monetary, socioeconomic, and cultural value, to Google and Microsoft commercially, and to users globally, compared to the cost or sales value that the original data collectors may have considered acceptable at the time of collection. The future value of information — all information, not just GI — is what underpins the whole industry of data mining and allied technologies such as data warehousing, i.e., locating and using/reusing existing information in innovative ways.

In discussing the value of GI, one can also ask the question of value to whom — the data owner or data user or society as a whole? All have legitimate claims on wanting to know more about the value of GI. Society in this case comprising businesses, government, and citizens. Data owners in the commercial marketplace may take various steps to increase the monetary value of the GI they offer, e.g., by product differentiation and adding value. Commercial vendors also often attempt to increase the net return (sales income) from their data assets through price differentiation, e.g., lowering the price for large-volume customers while charging a higher price for one-off use (more examples of price manipulation are discussed in Chapter 4). Yet, as already noted, users of, and uses for, GI vary so widely across business, government, and society that it is impossible to discuss the value of any one piece of GI for any one data user except in the context of the intended use. What is the untapped value of GI that has been collected for one purpose but not yet used for potentially myriad other purposes that may yield significant commercial and societal benefits? The very fact that the value is untapped means that we cannot assign a meaningful, defensible measure to that value, yet literally hundreds, even thousands of such cases exist if one simply takes the time to browse the stories, reports, anecdotes, or case studies in conference papers or scores of trade magazines both within and outside the GI industry.

2.2.2 *The relationship between cost and value*

Accept once again that all information has a cost. Geographic information has a range of direct costs, including collection, quality control, processing, storage, dissemination, advertising its existence, adding value, and use. No matter what value society as a whole assigns to certain types of GI or uses of GI, e.g., homeland security, disaster management, or monitoring climate change, it is not society that pays for GI, but rather individual people or organizations, public and private. These costs must be recovered by someone if information is to continue to be collected and used. If commercial information providers cannot recover these costs through efficient operation of the information market, they soon cease trading and the information disappears, i.e., it is no longer available to anyone for any purpose. If budgets of public GI holders (PGIHs) cannot sustain the cost of GI collection, dissemination, and use, then the information will disappear from the PGIH armory of tools that permit it to deliver efficient services to citizens.

Joffe and Bacastow (2005) propose that the cost or price that a user is willing to pay is a valid surrogate for perceived value of the GI being bought by a user, in a specific format, of specified quality, for a stated purpose, probably under legally binding contractual terms. The cost or price may vary depending upon different rights conferred to the user/consumer for different scenarios of use, e.g., own private use, use in one's own firm, use for clients, or use in a product or service for sale to a wider public. In the scenario proposed by Joffe and Bacastow, the user's cost will depend upon the data owner's policy, which can be represented in a cost matrix with parameters including "User Type by Data Access Right by Data Theme," and other costs may arise from the selection of different delivery methods or optional services. How public sector bodies charge for or recover such costs is a matter of considerable debate throughout the developed world, a debate now extending into developing nations as they build their National Spatial Data Infrastructures (NSDIs) with access to limited government budgets. These issues are discussed more fully in Chapters 3 and 4 on charging regimes and pricing issues, and in Chapter 6 on the role of GI value in cost–benefit analyses for SDI creation.

2.2.3 *Value determined by class of ownership, public vs. private*

Ownership of GI, and the motivation for collecting and selling or using that GI, highlights another aspect of the duality of value. Commercial vendors operating in the information market collect, process, and sell GI or GI-based services in order to earn an acceptable return on investment. Their primary concern is monetary (exchange) value from sale of the GI or related service. The added value to a user or to society as a whole is not as important as remaining in business. Public sector bodies that collect and use GI are concerned with doing so at the least cost, but the value of the GI or services that GI underpins is measured in terms of most efficient or enhanced service delivery to citizens, perhaps to other branches of government, and to society as a whole. Thus, whether GI is privately held, e.g., commercial sector GI (CSGI), or publicly held, e.g., public sector GI (PSGI), has direct impact on the value determination and the free or fee debate on charging.

Commercial sector GI has identifiable monetary value for its producers and vendors, e.g., look at published sales figures (Daratech, 2006) for the GI industry. CSGI has less quantifiable direct and indirect value to the economy and society resulting from the services offered using these data. PSGI has value to the government bodies that collect it initially to carry out their legally mandated governance functions more efficiently. While the cost of collecting and managing PSGI can be determined and, for the case of GI supply that is contracted out to third parties, can be very well defined, its value is not so easily calculated in financial terms, except to estimate the cost to government or society in terms of poorer quality governance or added cost of reduced efficiency if the data or service did not exist, i.e., the value of cost savings. Interestingly, when a public sector body buys (or licenses) GI from a commercial vendor, as is common practice today in many societies, the all-important monetary value to the CSGI vendor, who wishes to make a profit, is a cost to the PSGI buyer, for whom the true value may not even be quantifiable, monetarily, and if it is, the value may bear little relationship to the initial data cost.

2.2.4   *Summarizing issues in the GI value debate*

The relationship between cost and value is only one aspect of value of geographic information covered in this chapter, as there are other measures of value that have little relation to direct collection, processing, and dissemination costs. Cost and value will be further explored with regard to the information value chain for geographic information, considering that more than one type of value chain may apply. Changing information policies can alter the value of GI, reducing potential financial value for some data owners, both in private industry and for public bodies, while increasing value to others, or perhaps to society as a whole. For example, a policy change forcing cheaper, wider access and more liberal exploitation rights to public sector GI can make redundant or reduce the market value of some existing value-added services offered by commercial data providers prior to the policy change, yet create new value-adding actors in the industry, or permit easier access by citizens' groups to GI of value in achieving their goals.

In following the various arguments and insights into value of GI dealing with pricing and charging regimes and access issues in the remainder of this chapter and in Chapters 3 and 4, remember these basic points:

- Everyone, whether person or organization, is a user of information, which is at the heart of the information society and underpins the evolving knowledge society and knowledge economy.
- Geographic information manifests itself in many different forms and formats, for myriad uses, often in combination with other nongeographic information.
- The location attribute that defines information as geographic is only one of many attributes for that information, each of which has its own unique impact on information value.
- The value of information varies with time and according to different uses.
- All information has a range of costs associated with it, which must be covered by someone, although cost recovery alone is not the only measure of value.
- Different information value chains may apply to different stakeholders, and information policy at the national level or within organizations can affect the value chain.

Understanding the value of a good, including information goods, is essential in addressing the issues of pricing or charging for a good, whether in the private or public sectors. Pricing, charging, and access issues are covered in Chapters 3 and 4 and are included here only where they affect the value debate. This chapter also examines the claim by geographic information proponents from industry and government that GI is of special importance for society and the economy because it underpins most other information. This claim has a direct impact on how GI is valued in society, especially geographic information generated by government, i.e., public sector GI (PSGI).

Attempting to define the value of geographic information requires introducing several concepts dealing with value theory, the nature of information, and the value of information generally; valuing intangible assets; deciding which type of value is important, e.g., financial (monetary, exchange) value, economic value, social or cultural value; and investigating if there is a separate, specific value to the geographic component (the location or place attribute) of what is called geographic information. As indicated in this introduction, unlike the value of most physical goods, the value of a specific piece of information may vary greatly with time, quality, provenance, intended purpose of use, and even with how that information is recorded, stored, or disseminated. Let us look first at theories underpinning the concept of value itself.

2.3   *Value theory*

Value theory is a concept normally associated with decision theory that "strives to evaluate relative utilities of simple and mixed parameters which can be used to describe outcomes" (Anon., 2003). According to several experts' contributions to Wikipedia, value theories try to explain why and how people place positive or negative values on things (goods) or concepts, and the reasoning behind their evaluations. Value theories tend to differentiate between moral goods, i.e., those relating to conduct of persons or organizations, and natural goods, i.e., objects. Yet information is the sort of hybrid good that can be treated as a natural good, e.g., as an information product, such as a book or map, and as a moral good, e.g., if information is used to praise someone or enable creation of a public good, or misused to defame a person or pervert the course of justice. Can value theory help explain why valuing information is so problematic?

Economists propose that goods are sought in marketplaces and that consumers' choices and willingness to pay set the value for goods. Ethicists speak of intrinsic and instrumental goods, the former being of value by themselves and the latter of value in getting something else that may be of intrinsic value. However, since information goods can be both intrinsic and instrumental, this does not advance our understanding of the value of information considerably. Information as a commodity is presumed to have a value, an exchange value, a use value, and a price. Exchange value of a commodity is not necessarily the same as its price or the monetary value for which the commodity will be exchanged between vendor and purchaser, but represents rather what quantity of other commodities might be exchanged in the trade. However, since most of us today do not engage in barter trade when acquiring goods or services, exchange value is probably best thought of as monetary value based on the purchaser's willingness to pay for that good or service.

The link between use value and utility is explored by both philosophers and economists, from as far back as Aristotle. Since the utility of something to someone else, whether a product or service, depends upon many variables, the differences between use value, exchange value, and price can be considerable. A cheap hammer used to smash a window to allow your escape from a burning room has a utility value very much greater than the cost of the hammer. In the same way, the marginal cost of a single piece of information that permits you to complete a necessary job on time and more efficiently, advances your career, or saves your life — and it might be the same piece of information — bears little relation to the initial price you may have paid to gain access to that piece of information. We conclude that it is the very nature of information that prevents one from assigning a single value in any of the terms or parameters put forward in value theory. Rather, the same information can have a price and user willing to pay, which comprises the exchange value in modern society that satisfies a vendor or producer, i.e., permits the information product or service to remain available due to market demand. The same information can have a use value or utility that far exceeds the exchange value, depending upon factors too numerous to list and that vary across use, user, circumstances, and time. Yet both exchange value (price and willingness to pay) and use value constitute the true value of information.

2.4   The information market and the information economy

Shapiro and Varian, in their bestseller *Information Rules*, define information as "anything that can be digitized — encoded as a stream of bits," and "information goods" are the products made available based on such information, including databases, books, movies, or web pages (Shapiro and Varian, 1999, p. 3). Unlike physical goods, information is expensive to produce and inexpensive to reproduce, i.e., information goods have high fixed (sunk) costs and low marginal (reproduction) costs. Using today's information and communications technology (ICT), information goods also exhibit low, sometimes negligible, dissemination costs. For this reason, "cost based pricing just doesn't work … you must price your information goods according to consumer value, not according to production cost" (PIRA, 2000, p. 3). While this may be excellent advice for commercial vendors in the information marketplace, it has little relevance for government departments who are required to make their information resources available at no cost, cost of dissemination only, or some other artificially determined low cost that may bear no relationship to actual production cost. Information is also considered to be nonrival and nonappropriable and tends to exhibit high elasticity of demand (Pluijmers and Weiss, 2001).

Shapiro and Varian contend that only two models exist in a sustainable information market, i.e., the dominant firm and the differentiated product markets (Shapiro and Varian, 1999, p. 25), although combinations of the two also occur. The dominant firm includes monopolist data or service suppliers, both public and private, e.g., firms or agencies whose data must be used for legal purposes or that have inherited a historical monopoly on data supply for historical purposes, such as some national mapping agencies, hydrographic offices, census bureaus, or national statistical offices. Dominant firms exist in the marketplace for space-based imagery, due to the small number of data providers caused by the high cost of entry into this image collection business, typically hundreds of millions of dollars or euro to build, launch, and operate even a single remote sensing platform in space. Yet this special geographic information marketplace is also a differentiated market due to the different types of imaging sensors available on different platforms, differences in resolution or spectral coverage, periodicity (repeat passages over the same section of the earth), ability to penetrate clouds (radar vs. visible spectrum sensors), etc. Total dominance of the space-based imagery marketplace by a few firms is thwarted due to the number of government-owned and -operated services in this sector, several of which provide imaging products at cost of dissemination only, such as those operated by the U.S. government for which federal data are freely available due to national legislation, or to meet other national socioeconomic goals, for example, in Canada and India.

2.4.1   Information as an intangible asset

Since the bulk of the information industry comprises intangible assets, much has been written about the value of such nonphysical assets; for example, patents that protect manufacturing methods and ingredients for valuable pharmaceuticals or other machine inventions, or copyright or database protection for other forms of information, from books to movies to important national GI data sets. How do you assign value to intangible assets? Many organizations also undervalue information developed for internal use, which studies have shown may be one of their most important assets, even though it is exceptionally difficult to assign a numerical, monetary value to such information in order to include it on a corporate balance sheet.

There is another side to valuing an intangible such as information. For example, what is the value of information that helps reduce traffic accidents or deaths? Within Europe, the socioeconomic cost of road deaths and injuries was estimated at 200 billion euro per year by one study (RoadPeace, 2003), while "the most precise estimations of the total socio-economic costs of road accidents in the EU (including estimates for under-reporting of non-fatal accidents) exceed 160 billion euro annually, which is almost 2% of GDP; whereas attributing an economic cost to road fatalities and damages shows that the cost of preventing accidents is far less than the economic cost of crashes" (European Parliament, 2000, p. 6). But how does one assign specific value to specific GI that might be useful in reducing road deaths or injuries? Excessive speed and careless driving are responsible for many accidents and resulting deaths. So how can road or traffic-related GI help prevent these two exceedingly bad habits of many drivers? And if prevention cannot be ascertained, then how can you assign further value to the GI based on deaths prevented? If the principal causes of road death and injury "largely remain the same: speed and alcohol and non-wearing of restraints" (NZ Police, 2004, p. 15), then how would GI help prevent this and what is the added value (socioeconomic) of such GI? Another issue is to whom would you assign the value and benefit of information that helped reduce traffic injury or death — the government, the insurance companies, the citizen, all of these? What if the information is provided, but then has no impact in many cases because it is ignored or otherwise not applied? This reinforces the premise stated at the beginning of this chapter: information is only of value if used.

2.4.2 The role of technology and infrastructure

In the digital age, the information economy is driven as much by information and communications technology (ICT), and the infrastructure that underpins ICT, as by the information products on offer. Advances in technology have a direct impact on the value of many information products and services; e.g., the ability to distribute large volumes of data, on demand, in a format that the user can query or integrate directly into other products or services, using appropriate software. In this context, the value of the Web is widely recognized as the medium by which digital information goods can be disseminated at low cost, globally. Some of the current technology trends that will have direct or indirect impacts on wider access to and use of GI include:

- Increased computer power at decreasing cost (per storage byte, per instruction step), due to ever faster processors and storage technologies with ever larger storage capacity at continually reduced cost per byte
- More powerful platforms and infrastructures to access information, including multifunctional portable devices that combine the functionality of mobile phones, personal digital assistants (PDAs), and location-based devices (GPS enabling technology)
- Wider use of remote sensing devices for collecting GI and greater availability of software to process those data into formats for many different uses; e.g., very high resolution data becoming available from commercial satellite imagery providers coupled with sophisticated imagery analysis capability at the desktop
- Greater positional accuracy for location information both inside and outside buildings and in built-up areas, e.g., 2 to 3 cm accuracy with combined differential GPS and Galileo system
- Continued advances in integration of data collection and management systems, on land and at sea, that will decrease the cost of spatial data collection
- New initiatives in community-based mapping, in which members of the public collect their own spatial data using a variety of techniques, from handheld data collection to imagery interpretation, then make this available without intellectual property restrictions, typically via the Web, e.g., OpenStreetMap
- Advances in microelectronics and battery (power) technologies leading to ever smaller, more portable, and more powerful devices for location-based applications
- Growing use of real-time location data for personal navigation, in-car navigation, ship navigation (electronic chart display systems, or ECDISs), and aircraft navigation
- More and better integration of sound and visual data for delivering multimedia content to location-based platforms (especially in-car or handheld)

with content relevant to the receiver's location and convergence within the ICT and information content creation and delivery industries

- Advances in artificial intelligence (AI) and expert systems to open up spatial data portrayal and analysis capabilities to nonexpert users
- Evolution of Web portal technology and further development of the Semantic Web, driven by the World Wide Web Consortium (W3C)
- Wider integration and convergence of intellectual property rights (IPR) and digital rights management (DRM) technology within and across content types and information sectors; increased use of click-use or click-through licenses that permit rapid and legal access to large volumes of data that are not otherwise available for free
- Spread of broadband telecommunications capabilities to all users throughout more communities, both hardwired and wireless

These technologies have increased the value of digital information generally, whether private or public, and certainly for geographic information products and services. Increased value of an information good due to technology and infrastructure, compared to historical analogue means of production and distribution, may not necessarily result in increased price to the consumer. The sunk (fixed) cost for creating GI products or services remains the same as before, but a much wider market may now be reached for both promotion and sales, at much reduced cost to the data or service provider.

2.5   *The value chain*

The value chain is defined as the set of value-adding activities an organization performs in creating and distributing goods and services, including direct activities such as production and sales, and indirect activities such as managing human resources and providing finance. In Porter's (1985) classic production value chain, shown in Figure 2.1, as applied to manufacturing enterprises, goods progress from raw materials to finished products via a number of stages, during each of which new value is added to the original

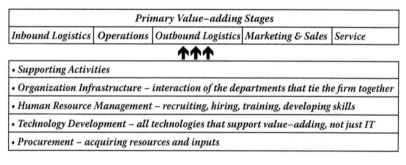

Primary Value–adding Stages				
Inbound Logistics	Operations	Outbound Logistics	Marketing & Sales	Service

⬆⬆⬆

• Supporting Activities
• Organization Infrastructure – interaction of the departments that tie the firm together
• Human Resource Management – recruiting, hiring, training, developing skills
• Technology Development – all technologies that support value–adding, not just IT
• Procurement – acquiring resources and inputs

Figure 2.1 The value chain according to Porter. (Adapted from Porter, M.E., *Competitive Advantage*, The Free Press, New York, 1985.)

input by various activities. If the value or price of the outputs at any stage is higher than the value or cost of inputs to that stage, then value has been added, resulting in a profit margin earned within that stage. The sum of all such margins, at the end of the chain, equals the total value added.

There may be hundreds of activities performed at each stage of the value chain shown in Figure 2.1, and any one may impact upon, or depend upon, other activities not only within that stage but at other stages. Value chain analysis is the systematic approach to examining the development of com-petitive advantage achieved when an organization executes the activities in its value chain more efficiently or more cheaply than the competition. The value chain is a useful way to identify, monitor, and judge performance of core competencies in both supporting and primary value-adding activities that lead to a competitive edge, i.e., creating a cost advantage over competi-tors, or a differentiation advantage (NetMBA.com, 2007). Having defined the value chain for a product or service, the organization can assign costs to the activities along the chain. A cost advantage is created by reducing cost of specific activities or by reconfiguring the value chain, i.e., redefining pro-cesses, marketing channels, pricing strategies, etc. According to Porter (1985), ten cost drivers are identified for the value chain activities, which, if better controlled than competitors', can lead to a cost advantage. The differentiation advantage arises from uniqueness in any part of the value chain, e.g., inputs not readily available to competitors or unique distribution channels, policies, or regulatory environments. Some of the nine uniqueness drivers that Porter identifies are also cost drivers. Differentiation may result in greater costs, for example, creating or expanding a unique, high-technology distribution chain. But if the associated costs add value that competitors cannot match, then the resulting total added value should be greater than if the differenti-ating activity was not implemented.

2.5.1   *The information value chain*

The value chain concept for enterprises producing goods or providing ser-vices has been extended into the information market via various proposals for an information value chain, i.e., adding value to information by various activities as it progresses from raw data to a new form of information or information service. Information and communications technologies (ICT) have a direct impact on virtually all the activities in the information value chain, by the very nature of information collection, processing, and dissemi-nation activities.

Does GI adhere to value chain concepts for determining the value of infor-mation, especially in relation to similar information, e.g., scientific, technical, and medical (STM) information? Since an estimated 80% of all government information has a geographic component (FGDC, 2004), what are the similar-ities and dissimilarities between private sector and public sector GI regard-ing perceived value, based on the many criteria that determine value? What happens in the value chain when private firms exploit public sector GI or when GI produced by the private sector becomes public, i.e., when governments outsource data collection to the private sector? Do access, exploitation, and intellectual property rights (IPR) impact on the value of public sector GI any more so than they do on the private sector? What does the term *value-added GI* mean — does GI itself have value added, or only the services that use GI?

These are some of the questions that need exploring in regard to GI and the information value chain. For example, when value is added to an initial piece of GI, then this new GI has its own unique value, distinct from that of the original information. Wehn de Montalvo et al. (2004) point out that "location-based mobile services will come to be fully integrated and seamlessly available to end-users seeking localised and customized content, which has value-adding implications for the location-aware component of the content." The U.S. Office of Management and Budget (OMB, 2005, p. C-1) defines an information value chain model as the "set of artifacts within the (enterprise) describing how the enterprise converts its data into useful information."

2.5.2   Which information value chain for GI?

We propose that the value chain perceived by public sector GI owners (government agencies) who collect and use such GI for legally-mandated purposes relating to governance of society differs from the value chain for commercial actors in the information market. Does the PSGI manager actually care about the value chain, in the same sense as a commercial information product or service provider, even though both types of owner/user typically do add value to information between collection and use? Many authors have proposed different information value chains for different types of information and from different viewpoints. Spataro and Crow (2002) propose the five-stage value chain shown in Figure 2.2.

Oelschlager (2004) defines the information value chain in Figure 2.3 in terms of enterprise-wide information integration that converts unstructured data arising from business processes to "actionable information."

Phillips (2001) proposes a management information value chain (MIVC) based on six types of value-enhancing activity, as shown in Figure 2.4. The goal of the MIVC is conversion of raw data into useful information that is then acted upon by management, contributing to corporate value or enhanced organizational efficiency.

MIVC is based on two assumptions. First, management information systems provide information to enable better decision making. Second, the value of such information equals increased profitability or greater organizational efficiency due to better decisions being made. The value added to the raw data by the intermediate activities, post-acquisition until final use, is measured by the extent to which each activity contributes to the main goal. Initial transformation activities include aggregating and filtering raw data,

Stage 1	Stage 2	Stage 3	Stage 4	Stage 5
Create	*Manage*	*Integrate*	*Transact*	*Distribute*
Collect, organize, add context	*Store, prepare for multiple uses according to an information model*	*Locate and aggregate from multiple sources; produce information intelligence*	*Mechanisms for realizing value and content monetarization*	*Deliver content to end–users in a suitable information package*

Figure 2.2   The content management information value chain, adapted from Spataro and Crow (2002).

and integrating multiple data sources. Dissemination involves getting the right information to the right people when needed, which includes determining who needs what and in what format. Modeling and presentation actions then transform the integrated information into the necessary format for immediate use to different levels of decision maker. In the final two stages of the MIVC, IT-oriented activity is replaced by humans making and acting on decisions based on the information presented to them.

The MIVC offers a good candidate value chain for GI because a great deal of the GI collected by government and private industry is used to help make

Stage 1	Stage 2	Stage 3	Stage 4	Stage 5	Stage 6
Unstructured Data	*Structured Data*	*Contextual Information*	*Business Information*	*Knowledge*	*Active Insight*

Figure 2.3   Information value chain according to Oelschlager (2004).

Stage 1	Stage 2	Stage 3	Stage 4	Stage 5	Stage 6
Data Acquisition	*Initial Transformation*	*Dissemination*	*Modelling Tools and Presentation*	*Decisions*	*Actions*
IT Actions				**Management Activities**	

Increasing Information

Figure 2.4   MIVC, adapted from Phillips (2001).

decisions in which the location attribute is an important part of the deci-sion-making process. There are quite specific, often expensive and complex, activities taking place in the data acquisition, transformation, modeling, and dissemination stages that are unique to GI compared to other forms of infor-mation to which the MIVC also applies.

From the viewpoint of the raw data provider, note that the new informa-tion created at each stage in the value chain is not the same as the data or information in the prior stage. In other words, adding value to raw data, for example, a road centerline, by integrating it with other sources of infor-mation, attributes, models, and dissemination technology to provide, say, a road navigation service, does not change the inherent value of the road centerline data. The navigation service provider's willingness to pay for the same or similar data elsewhere in the road system remains the same. New value is created at each stage by activities that require expenditure of resources (money, human capital, infrastructure). Such expenditure should not be undertaken unless the result is information of value greater than the combined cost of the value-adding activities in each stage and the cost of the information as it entered that stage.

2.6   Different components of value for GI

Understanding value of GI requires a closer look at the relationships between data and information, attributes and context, timeliness and quality, and other factors that can add value to raw geographic data. Data represent facts or features about the real world. For example, a single point, specified in some meaningful spatial reference system, perhaps denoting a specific location on a road centerline or a property boundary, is a piece of data. But that datum, perhaps a grid reference number or lat/long pair, means little to anyone, and has little value, until more attributes are added to the overall information package. Additional information is needed to add meaningful context to that point, i.e., its definition as part of a road centerline or a boundary line, rather than simply some random point on the surface of the earth. Additional attri-butes add further contextual content to the original data point, for exam-ple, something about its accuracy, precision, provenance (who surveyed the point), history (when was the data collected, validated, updated), or method by which it was measured. All of these additional information elements add value to the raw data, resulting in a more robust information package that can be used in a range of contexts.

2.6.1   Value of the location attribute in GI

Spataro and Crow (2002) define data as "transaction-based information," while content is "context-sensitive information." In their information model, raw data assume a new value, as context-sensitive information, due to struc-ture created by wrapping an information package in a metadata wrapper, resulting in a "content component." The added value created by metadata is discussed more completely in a later section. For geographic information, the location attribute provides spatial context to the other attributes in the information package, thus increasing the value of the data for applications where spatial awareness is key.

Much geographic information is said to have special value as an underpinning framework for other information and services; e.g., location information as an attribute of other important information, such as occurrences of disease or storm paths or road traffic accident data. The disease, meteorology, or accident information has many important attributes other than just location. These nongeographic attributes have value in their own right; e.g., information specific to the type of diseases and their impact on society plus cost of prevention or remediation, or severity of storms and degree of damage they may inflict on society and cost to insurers or property owners. Is it possible to set a value on just the geographic (location-based) component of such data? The mixture of attributes, geographic and nongeographic, that typifies much geographic information further complicates the process of setting a single value on such information.

For example, information on crop growth rates for a single farm, aggregated to cover an entire region, can be extremely valuable in regard to the level of crop subsidy likely to be paid from government coffers to a single farmer or for the whole region. The same information is important in regard to the regional *in situ* capacity for crop processing and distribution services that may be needed, with ramifications for local employment levels and purchase of local supporting services. The location attribute alone for the crop growth rate data for a single data point in a farmer's field, from which all aggregated data is derived, may be of little value to a regional planner, but of great value to a farmer who can act on it by, for example, applying fertilizer at different rates across a single field, thus increasing yields, using geographic information system (GIS)-based precision farming techniques. The information contained in the location attribute is collected simultaneously with the other crop data, yet for the most part only has value if used in relation to one or more of the other attributes of the crop data, e.g., type and variety of crop, local plant height or density, or grain kernel size. This example also demonstrates the potential difference in value of geographic information based on its granularity; e.g., point data (location where a reading was taken) has one value, field- or farm-wide data (aggregated point data) has another value, and regional data (data aggregated across many farms) has yet another value, each value dependent upon the intended use and the perceived value of the information package to the user.

2.6.2   *Time dependency value of GI*

The value of a television program listing degrades rapidly with time. Once a program has been broadcast, the schedule listing that program, whether printed or online, is of no further value to most television viewers. However, it may be of value to researchers into the history of media or broadcasting, for biographies relating to script writers, producers, actors, and actresses, or even for legal purposes. Yet one week prior to the airing of the program, the schedule information was of sufficient value for a customer to be willing to pay for a program listing or for an intermediary, such as a newspaper, to pay a listing service for the information to include in its daily issues.

Similarly for GI, traffic congestion information, with an undisputed geographic (location) element, is of high value as the congestion occurs, e.g., to issue warnings to motorists of delays or obstructions, or for emergency services to react to accidents. It is of reduced value once the congestion has cleared, and of much reduced value 24 hours later, except to provide a historical or statistical picture of congestion black spots in the road transport network. Another example of the time-related value of GI is meteorological data used to prepare weather forecasts that underpin or influence myriad decisions at private, commercial, and government levels. Again, such information rapidly loses value, except for historical, analytical purposes, e.g., preparing weather-related statistics for regions or longer-term climate change research. Information from a decadal population census "declines in value as it ages in the 10 years between censuses," but the value rises again following the next census, when it forms the benchmark against which change is determined over the preceding decade (Longley et al., 2001, p. 376).

Therefore, the value of certain types of geographic information may depend on whether it is real-time data (happening now), near real time (will affect near-future events), relatively invariant (location of a building), or historical. The decision by the data owner as to when to make that information available, to whom, and at what price is a marketing decision that assumes different values for the same data, depending upon user needs and perceptions of that value and intended uses.

2.6.3 Value determined by cost savings

One of the basic principles of most spatial data infrastructure (SDI) strategies is that data should be collected at one level of government, i.e., the most appropriate level, and then shared among all levels where possible. The importance of this concept in developing a national SDI is covered more thoroughly in Chapter 6. The rationale for this goal is to save the cost of duplication in collecting data. As an example, Cobb (2002) quotes studies conducted by Bhagwat and Ipe (2000, p. 21) for the state of Kentucky geological mapping program, in which they "determined the value of a 1:24,000-scale (7.5-minute) geologic quadrangle map to be $43,527. The respondents … said they saved this amount, on the average, because the maps were already available and therefore they did not have to collect the data themselves." In a review of the MetroGIS urban system in Minnesota, cost savings were noted in that "the value of the regional datasets was not in the data but in elimination of the need to individually internalize the costs to integrate/merge the data received from multiple counties" (MetroGIS, 1999). Numerous other examples exist where cost savings are the main determinant of benefit in cost–benefit studies, which will be explored more in Chapter 6.

2.6.4 Adding value via information management techniques and tools

The value of GI can be increased based on how it is recorded, i.e., the physical medium by which it is captured and represented, data formats used, and metadata made available. These have a direct impact on how the data can be disseminated and incorporated with other data sources, and at what cost. In most cases, it is far easier to disseminate and add value to digital data than a raster (map) image — hence the almost universal drive to digitize currently analogue GI data holdings. The UK Advisory Panel on Public Sector Information (APPSI) recognized that use of document and knowledge management systems also plays a role in increasing the value of information. "In our view, the value of public sector information will only be realised and exploited when fairly advanced systems of each kind are in place," and "full exploitation of public sector information will depend on the presence of advanced systems ... for identifying and making available information in electronic form" (APPSI, 2006, p. 20).

GI available in open-source, standardized markup format may have higher value due mainly to the markup format, especially in regard to provision for Web services. "The main criticism of data was not access or quantity, but availability in the right form" (APPSI, 2005, p. 6). Thus, getting GI into the right form can alone add value, without otherwise enhancing the information value tied up in the different attributes themselves. Similarly, adding adequate, standardized, readily accessible metadata to data sets can increase the overall value of the data set because it can be more easily located using appropriate search tools, more easily understood, and thus perhaps used more wisely and misused less often.

2.6.5 Value due to legal or other mandatory use requirements

In many legal jurisdictions, certain information is given an official or legal status for certain types of transactions. One of the most common examples relating to geographic information is the boundary data in cadastral or similar land registration systems, whether urban or rural. In this perverse world, even if the legal boundary line drawn on a map by hand many decades previously is not exactly reconcilable with the actual boundary on the ground between properties, the cadastral map typically takes legal precedent. Data from officially recognized agencies, typically a national mapping agency, land registry, or official address or gazetteer owner, must often be used in other civil applications as well, sometimes conferring a monopolistic, or near-monopolistic, position in society regarding such data. Even if identical data are collected by a third party, perhaps by the same contractor who collected the data originally for the government agency under a managed subcontract, if the law or internal regulation or standard operating procedure requires that the official data must come from a specific supplier, then that supplier enjoys a value advantage unrelated to the quality, timeliness, etc., of the data in question. Some regimes recognize that not all official data may be available from the preferred or mandated government agency and make allowances for nonofficial data to be used in its place until such time as official data are available and registered — for example, the regime operated in Catalunya, under its law concerning the use of geographic and cartographic information for this autonomous region (Generalitat Catalunya, 2005).

2.6.6   Value due to network effects

Some information goods have added value simply because they are used by large numbers of people, i.e., they become a standard by which other, similar or even identical information is judged. In the U.K., this is the position enjoyed by the premier mapping product of the Ordnance Survey, with its MasterMap® product, a digital representation of the real world containing more than 450 million uniquely identified geographic features, updated as many as 50,000 times per day, providing a "consistent framework for the referencing of geographic information in Great Britain" (Ordnance Survey, 2007). MasterMap is so widely used in the U.K. that it enjoys a considerable network effect, even though some users complain about the cost and that the information should be freely available as an important part of the U.K.'s public sector information asset.

2.6.7   Value due to quality of an information resource

What determines quality, especially in regard to different consumers and uses for geographic information? Assume first that we are talking about a data set or service that suits a specific type of consumer and intended purpose of use, e.g., the scale is appropriate, the format is acceptable, etc. Then typical quality issues include completeness of the data set, timeliness of the information, and provenance or reputation of the information provider, all of which can add or detract from the value as perceived by the consumer.

Completeness or comprehensiveness of content varies greatly across GI products and also with users' requirements. Especially if one is paying for a mapping product, there is little sense in the consumer paying for completeness or accuracy that is not needed. On the other hand, certain data sets need to be as complete and accurate as possible, for example, the road infrastructure data that underpin in-car navigation systems, where consumers can become quite irate when whole streets seem to be missing —

especially the street they find themselves driving down in a strange city. Another aspect of completeness is timeliness or frequency of update, which is often the most expensive part of data set cost overall, as evidenced by the amount of investment that goes into survey work by national mapping agencies and companies like Tele Atlas or NAVTEQ, who "drive the roads in over 60 countries in order to keep their road infrastructure data set up to date" (NAVTEQ, 2007). Another aspect of timeliness is the frequency of coverage of parts of the earth by satellite-based remote sensing platforms. No matter how superb the observation instrument may be, if it is not pointed at the area of concern to the customer when the information is needed, for example, to monitor farm crops, oil spills, or other natural disasters, then the high-precision images are of little value.

The reputation of the data provider can also add value. Some information is perceived by consumers to be of higher value because of the reputation of the data provider; e.g., consumers of financial information are more likely to value financial news from an organization such as the *Wall Street Journal* than from an unknown Web-based financial news start-up company. Unlike most physical goods, information is an "experience good" (OXERA, 1999). In other words, to know if it is of value, you must have access to it first. Yet most consumers are reluctant to pay for a good not knowing if it is suitable for their requirements prior to payment, unless there is some form of money-back guarantee. Such guarantees raise a new problem for information providers in regard to information goods, since, once consumed, the information content cannot be returned in the traditional sense of the word, as would apply to a physical product.

If past personal experience or recommendations from other consumers or professional experts indicate that a certain data provider has an excellent reputation for the type of information good being purchased, then consumers are more likely to pay the price asked, even if this is at a premium compared to other data providers, because of the added value they perceive due to the provider's reputation.

Finally, conveying information on the quality of the GI on offer can add value, especially for new or uninformed consumers of that data. Value of an information good or service, as perceived by consumers, is an important element in setting acceptable prices, and thus determining the level of remuneration available to a data or service provider. What happens to this element of the value consideration when the consumer has little experience with the type of information on offer? For example, a local government body acquiring geographic information for use in a GIS to help manage a street maintenance program will almost certainly have access to experts with knowledge of the best data supplier for their needs, whether this is a national mapping agency, utility company, or similar. But laypersons accessing Google Earth may not understand the importance of completeness of coverage, or precision, accuracy, or timeliness of the imagery used to underpin Google Earth. In ignorance of the value of high-quality data, how many Google Earth users would be willing to pay premium prices for access to better-quality data in their Google Earth experience?

2.7   Value of geographic information to economies and society

Geographic information has commercial, economic, and socioeconomic value, the latter not always being easy to define. Even for commercial value, different market studies define the geographic industry differently, so it can be difficult to compare figures between studies and over years. For example, is a GI data set that is delivered as part of a product valued as data or part of the GIS application itself, in which case, where in the market figures would this contribution to market value appear? Even more taxing is the question of just how much GI contributes to an economy in total, taking into account both direct and indirect effects; i.e., quantifiable additions from added employment, taxes generated from sales and services, etc., vs. cost savings to government and businesses from wider use of GI leading to more efficient operations.

2.7.1   Commercial value of GI

In 2004, according to Daratech (2006), global geospatial data revenue was U.S.$677 million, one quarter of total forecast global market revenue of U.S.$2.82 billion for the geospatial technology industry comprising software, data, services, and hardware. The commercial value of data was second only to software and well ahead of services and hardware sales in the same period. The 17% increase in market size experienced in 2005, to U.S.$3.3 billion, was led by growth in data products. Government was the largest single sector, accounting for one-third of total revenue in the industry, forecast to produce revenues of U.S.$3.6 billion in 2006 across all sectors. However, commercial data sales revenues tell only part of the story of the value of geographic information (GI) to industry, governments, and society as a whole.

2.7.2   Economic value of GI

Attempts have been made to define a value for GI as a class of information, focusing on its place- or location-based attributes, in relation to whole economies or as a component of the total market for all types of information, especially for public sector information, i.e., information collected by public bodies (PIRA, 2000; OXERA, 1999). Such analyses typically focus on added value for GI due to the ease of access to the information and ability of others to easily acquire and exploit public sector GI at minimal cost, preferably cost of dissemination only, as is the rule for much federal data in the U.S. A study by Pluijmers and Weiss (2001, p. 26) focuses on "maximizing economic and social benefit from the dissemination of information and data already acknowledged by governments as not confidential, and ... fair terms for commercialisation of government data and competition with the private sector."

In the U.K., the Office of Fair Trading (OFT) launched a study explicitly acknowledging the value of public sector information within the economy, even if that value could not yet be scientifically or numerically quantified (OFT, 2006). The study looks at how PSI is turned into value-added information; how the pricing of PSI and access to it affects competition between public sector bodies and private sector vendors; and the effectiveness of existing guidance and laws, specifically the U.K.'s Re-Use of Public Sector Information Regulations 2005, which implemented the 2003 EU directive on reuse of PSI (APPSI, 2006, p. 4). The study concluded that improving access to and exploitation of PSI in the U.K. could "double in terms of the value it (PSI) contributes to the UK economy to a figure of £1 billion annually."

2.7.3   Socioeconomic value of GI

Socioeconomics is the study of the relationship between economic activity and social life, examining economic impacts of social activity and social impacts of economic activity. While economists tend to look at economic impacts in financial terms, socioeconomists typically focus on the social impact of some sort of economic change, such as advances in information and communications technology (ICT), changes in intellectual property (IP) law, or changes in government information access law or privacy laws. The social effects of such changes can be far-reaching and unforeseen. Similarly, social activities, such as file sharing over the Web, can have important impacts, both positive and negative, on information economics, whether done illegally or by using legal file-sharing services.

Socioeconomic impacts "may affect patterns of consumption, the distribution of incomes and wealth, the way in which people behave (both in terms of purchase decisions and the way in which they choose to spend their time), and the overall quality of life" (Wikipedia, 2007). In analyzing these impacts, socioeconomists use metrics such as improvements in literacy, employment, and shifts in employment between sectors, which are relevant to wider access to and use of information, taking into account negativities of the digital divide dilemma. Other socioeconomic factors, more difficult to measure, that are impacted by the information industry include freedom of association (online censorship), ability to participate in society (exclusion from the digital exchange), and fears over personal safety, e.g., online stalking, cybercrime, and online child pornography.

As defined above, most, if not all, public sector GI has socioeconomic value. It is necessary for governance of society, and that is its primary legitimate use and rationale for collection using taxpayers' money. Commercial GI also has socioeconomic value or no one would pay for the products and services on offer. Many of the value-added information products and services that industry produces are extremely important to society; e.g., on-board vehicle navigation systems and weather forecasting services are used by private individuals, businesses, and government agencies for a wide range of purposes, from the mundane to those that save property and life.

Thus, the main distinction between commercial value of GI and value to society as a whole seems to be that commercial value is what drives industry to create GI products and services, whose success or failure depends upon who in society uses those services and how. Value to society as a whole can vary widely because the same GI product or service can often be used in many different ways, from entertainment or leisure activities to greater efficiency in business and government, to saving lives. These different values, as perceived by the buyer or consumer, may vary widely, leading to differences in commercial value to the vendor (sales price vs. production and sales cost), which will be explored in greater detail in Chapter 5 on pricing and charging considerations.

Another way of looking at the value of GI to society as a whole is to consider documented benefit–cost ratios for various types of GI. For example, geological maps are typically produced at a national level by government agencies supported almost entirely by public funding, rather than by the private sector, thus adding to the body of public sector GI available for use by all. Assessments in several U.S. states (Illinois, Virginia, Kentucky) reported benefit–cost ratios ranging from 5:1 to 54:1 for geologic mapping, and "given the myriad uses to which a map is put over its lifespan, the true benefit-to-cost ratio of geologic mapping must be greater still" (Utah Geological Survey, 2006).

2.7.4   Valuing the economic contribution of public sector GI

One relevant trend in Europe today regarding geospatial data within the information economy is the drive toward greater exploitation of public sector information (PSI). This is the focus of the EU directive on reuse of PSI, which came into effect on July 1, 2005. According to the PIRA study on the exploitation of European PSI (PIRA, 2000), investment value in European PSI was estimated at 9.5 billion euro per annum in 2000 and 19 billion euro per annum in the U.S. Investment value is defined by PIRA as the investment in the acquisition of PSI. PIRA defines economic value of PSI as "that part of national income attributable to industries and activities built on the exploitation of PSI ... the value added by PSI to the economy as a whole" (PIRA, 2000, p. 15). In Europe, they estimate that PSI economic value was 68 billion euro in 2000, compared to 750 billion euro in the U.S. PIRA also estimated that the GI industry accounted for 36 billion euro of this 68 billion euro total. One conclusion of the PIRA study was that the economic value of PSI was so much greater in the U.S. than in Europe, which at that time (EU had 15 member states) had roughly the same population and potential market size as the U.S., because of the much higher degree to which PSI was used within the information market, with value added by thousands of firms employing tens of thousands of skilled workers in the information sector.

The goal of the PIRA International Ltd. (2000) study was not to conduct a cost–benefit analysis (CBA) for SDI creation, but rather to examine market size for public sector information (PSI) in Europe, compare this to global competitors, e.g., the U.S., in the information marketplace, and to make recommendations as to how Europe could better its position in that market-place. The reason that we include the PIRA study in this report is to intro-duce the definition for value of information that was adopted by PIRA in conducting its study. Also, the study found that the value of the GI sector, at 38 billion euro, was the single largest sector for the projected European infor-mation market size for PSI of 68 billion euro, with the next nearest sector (economic and social data) reaching only 11.8 billion euro. By comparison, the value assigned to the U.S. information market was 750 billion euro in 1999. The discrepancy for the two regions of approximately the same popula-tion was ascribed to the open exploitation policy for most PSI in the U.S.

As to assigning value to information, PIRA's methodology differentiated between investment value and economic value. The former is what gov-ernments invest in acquiring PSI, while the latter represents the portion of national income "attributable to industries and activities built on the exploi-tation of PSI" (PIRA, 2000, p. 15), i.e., the value added by PSI to the economy as a whole. Economic value far surpassed investment value (an average fig-ure of 68 billion euro compared to 9.5 billion euro for investment), but the traditional source for economic value figures (national accounts information of traditional industries) is not available for the information marketplace. Hence, the first assumptions that creep into the analysis are that "estimates of the value added by users to PSI … provide figures for the economic value of PSI." Whereas investment value, relating directly to costs spent in acquir-ing PSI, was quite accurately estimated at 9.5 billion euro (of which, again, GI was the largest single sector at 37%), the economic value figure used is actu-ally a central estimate (not a simple average) based on a range of 28 to 134 bil-lion euro. As with cost–benefit ratios in the stratosphere, one also begins to question entire economic analysis reports built around assumptions leading to such widely varying values for one of the key components of the analysis, i.e., economic value.

A report by Oxford Economic Research Associates Ltd. published for the Ordnance Survey GB (OXERA, 1999) was commissioned to provide eviden-tial support for the importance of the role that OSGB, and geospatial data (topographic mapping, in this case), played in the economy of the U.K. as a whole. The economic value of OSGB as the primary map-producing agency in the U.K. was defined in the report as "the contribution which OS makes to the Great Britain economy as a producer of final and intermediate products and services, as a purchaser of intermediate products and services and … as the provider of geographic information (GI) in the national interest." Tell-ingly, the study also begins (paragraph 2, p. 1) with the warning that "mon-etary values provided are … broad indicators of the scale of the contribution of OS to Great Britain's economy. Given the lack of empirical evidence for a study of this kind ... conclusions are reached on the basis of both qualitative and quantitative assessments." This perhaps provides a good case for a cost-effectiveness analysis, as opposed to a traditional CBA.

As to methodology, the study categorizes benefits as gains of three types: increases in efficiency, increases in effectiveness, and provision of new products and services. These are achieved by reducing processing and search costs, reducing waste by better scheduling, reducing uncertainty for more efficient service delivery, and matching products and services to user needs. The report then assumes that "development of computer-based GIS ... has increased the efficiency and effectiveness with which GI is used throughout the economy." The study (p. 5) does acknowledge that there are many uses of GIS that generate significant benefits, not all of which are monetary, e.g., in health provision, social services, etc.

The OXERA study begins with the statement that in 1996, when Ordnance Survey operating costs were approximately U.K.£78 million, its "products and services contributed to 12 – 20% of gross value added (GVA) in the UK, equal to UK£79 to UK£136 billion GVA" (OXERA, 1999, p. 1). Even taking the lowest GVA figure, this figure could be misinterpreted as a raw benefit–cost ratio of 1000:1. However, the calculation is further explained by the following line: "this economic contribution of OS comes, in the main, through the use of OS products and services as a primary input into the production of several key sectors of the economy." Some in the GI industry, and those engaged in the policy and politics of SDIs, try to use the OXERA study as a proxy for a more traditional CBA, proving that almost any level of investment in an SDI is warranted. After all, who could argue with a benefit:cost ratio of 1000 to 1?

Sadly, this association between use of maps or other GI provided by OSGB and the value of the economy does not stand up to closer scrutiny, as it presumes that this GVA is possible only because the maps or GI exists, and that there are no competing alternatives that could deliver the same functionality except by use of topographic GI from OS. More to the point, as more than one study has warned, as soon as cost–benefit figures become too good to be true, it is time to start questioning the methodology, statistics, or analysis used in their calculation. Most decision makers responsible for significant investments in projects the size of a national SDI simply do not believe such apparently wonderful ratios, as they are so far out of the ordinary range typically encountered that they seem immediately suspicious, even if they are factual.

Geoscience Australia, in a 2002 submission to the Australian House of Representatives' inquiry into resources exploration impediments, highlighted the benefit–cost ratio of public provision of national- to regional-scale fundamental geoscientific data sets. The report contends that such data sets are an important foundation on which private sector mineral and petroleum extraction industries depend in conducting more focused geological surveys in an industry that accounted for AUS$55.3 billion to Australia's economy in 2001–2002. The publicly-available information provides this regional-scale, precompetitive knowledge base, online for free or at cost of distribution, to reduce risks in selecting areas for new exploration work by industry, thus "providing a competitive advantage for Australia" (Geoscience Australia, 2002, p. 16). One factor in Australia's position as a major minerals exporter has been "the greatly increased knowledge of the geology and resource potential ... resulting from systematic geological and geophysical surveys." The Australian model, a partnership in which the private sector invests in and performs exploration using knowledge created in the public sector, "is widely regarded as 'world's best practice.'" The report (p. 11) quotes exploration expenditure by private industry of AUS$5 (ranging from AUS$2.50 to AUS$10.00) for every one Australian dollar invested by government in precompetitive geoscientific data collection, resulting eventually in AUS$100 to AUS$150 of in-ground resources discovered, thus resulting in a benefit–ratio of 5:1 for industry investment based on the public sector GI, and 100:1 to 150:1 for additions to the proven mineral wealth of Australia.

2.7.5 Value of GI as underpinning for other information and services

To what extent is geographic information (GI) the fundamental hook on which other data, such as official statistics, can be structured, analyzed, and used? The Australia New Zealand Land Information Council (ANZLIC, 2005) states: "Australia's and New Zealand's economic growth, and social and environmental interests are underpinned by quality spatially referenced information. Note: 'quality spatially referenced data' means spatially referenced information that is current, complete, accurate, affordable, accessible and integratable." Concerning the value of GI for purposes other than that for which it was originally collected, the Utah Geological Survey (2006) offers examples of using geological maps to assess the adverse impact on land use practices such as dam failures and serious groundwater contamination "at countless thousands of sites across the nation." As to underpinning services, refer back to the reports in the previous section (p. 4) on GI underpinning economies generally.

2.7.6 Intangible benefits: value unquantifiable in monetary terms

Much information has a direct or indirect value for which it is exceptionally difficult to attach direct monetary, financial value. For example, what is the value of information that would help prevent a death on the roads or due to a natural disaster or health epidemic? Does the value depend solely upon how many lives may be saved and the accepted value of a single life? And who determines that value? Insurers and government agencies have both tried to assign a value to a human life, even developing formulas to be used in calculating the sum to be paid out (insurance) or to be acceptably spent in preventing a death (cost of road traffic schemes, road improvements, etc.). We will touch on this aspect of value in more detail in Chapter 6 in regard to the difficulties of assessing intangible value in cost–benefit studies for SDI implementation.

Cobb's report (2002) on the value of geologic maps in Kentucky quoted intangible benefits, including improved credibility of studies conducted using the unbiased information in maps prepared by scientists without vested interests, reduced time to complete projects, and the continued use of the maps for land, water, and environmental management, rather than minerals exploration, their original goal, 30 years after they were produced, by "so many diverse users that listing them all is almost impossible."

We looked generally at the cultural and political value of information in the opening sections of this chapter, after Bryson (2001), but how does one assign cultural value to geographic information. What is the value of historic maps and other GI used for interpreting and understanding history, the development of societies, tensions leading to wars, and the aftermath of wars? How is GI used to understand the interactions of peoples within their own societies, i.e., major displacements for environmental (natural disaster) or political (wars, discrimination) reasons, etc. — and what is the value of such GI? Or is the value in the application and not in the data? But if that is the case, then how do you value the data? These are almost metaphysical questions and considerations that will probably forever prevent assignment of any measurable value to such information.

2.8   The changing value of geographic information

Information policies, especially regarding access and reuse of GI, advances in information and communications technology (ICT), and evolution in the framework or infrastructure of the information market can all impact on the value of GI, potentially both reducing and increasing value to some sectors of society, including private industry. A separate issue is the value of geographic information systems (GISs), the tools that use geographic information and which have little or no value without access to GI.

2.8.1   Increasing the value of GI

There are several ways in which data owners or licensees can increase the value of digital information, including geographic information:

1. If you have not already done so, create and publish metadata, preferably using international standards, such as Dublin Core (ISO 8601) for discovery purposes, especially information that will help users — or in the future, automated search engines — identify appropriate uses for your information.

2. Use industry standard formats or markup languages, such as Geography Markup Language (GML), to make your data more accessible and easier to use in evolving new service architectures.
3. Reexamine your access and reuse policies to see if relaxation of restrictive policies could in fact generate wider use and greater benefits, especially socioeconomic benefits if you are a provider of public sector GI.
4. Adopt technology and policies that increase prospects for interoperability of your data sets with others.
5. Externalities that can lead to added value, i.e., events, changes, or evolution in the information market itself that could lead to increased value of your GI, consist of wider adoption of standardized digital rights management (DRM) technology, including the markup and automated management frameworks or infrastructure that remove fears by data owners over loss of control of their intellectual property.
6. "Data value can increase when users have the ability to see its potential when displayed with other available (atlas) layers" (O'Dea et al., 2004, p. 6).

2.8.2   *Restricting the value of GI*

If one accepts that the ultimate value of any GI will be derived from its widest possible use, then anything that prevents such wide use could be considered to reduce the overall value of geographic information. O'Dea et al. (2004), referencing Bartlett (1999), state that "lack of quality metadata can render specific datasets virtually useless due to the uncertainty of data quality and reliability." According to Joffe and Bacastow (2005, p. 8), "The current inability to confidently control the description, trading, protection, monitoring, and tracking of intellectual property rights has been a barrier to broader adoption of web-based geospatial data distribution. Therefore, a vast amount of public geodata remains unavailable." The implication is that uncertainty over intellectual property rights relating to GI actually does reduce its widest distribution and use, thus reducing its overall value to society.

Intellectual property rights (IPR) protect data owners via, for example, copyright and database protection rights set out in international conventions and implemented in national law in almost all countries. If IPR becomes overprotective, then enforcement can limit wider access and exploitation of geographic information, thus reducing the potential value of some GI to society as a whole, even though the financial value to the data-owning organization is protected. If IPR terms are too restrictive or too complex to be understood easily by humans, or by access negotiation software used by data grids, computational grids, or e-commerce systems in the near future, then IPR can work to the detriment of the data owner by preventing or reducing potentially valid sales. Ongoing standardization initiatives for geospatial data IPR, such as the Open Geospatial Consortium's GeoDRM (Geographic Digital Rights Management) open specification (Vowles, 2006), are attempting to overcome some of these potential barriers by providing a means to encode IPR and digital rights management information with the geospatial data itself.

Legally binding terms of use (ToU) are often incorporated in an online end-user license agreement (EULA), which the user enters into via various online license procedures, variously called click-use, click-to-use, or click-through licenses. Such legal instruments can be used to assign different end-user values to the same GI due to the terms and conditions imposed on the user or buyer, e.g., permitting for own use only or restricting various levels of redistribution or other forms of exploitation, with or without further value being added to the initial data. Experiences gained to date with this potential value-adding technology are summarized in an Open Geospatial Consortium (OGC) interoperability report (Wagner, 2006) that focuses on access control and ToU click-through IPR management.

Overzealous application of personal data privacy regulations can also restrict the value of information. For example, "as many epidemiological investigations and analyses have shown, the societal value of certain health data is potentially huge" (APPSI, 2006, p. 30), yet medical information "has always been a difficult area in which to fully exploit its intrinsic value." Epidemiological studies provide an excellent example of the vital importance of the location attribute of a piece of (health) information when underpinned by the geographic information needed to permit spatial analysis. Many of the other attributes of that same health data concern the patient only and are of little or no use (or value) to any one specific epidemiological study. Methodologies for extracting the useful attributes from the nonuseful, while still protecting patient confidentiality, add value to the data that would otherwise be restricted.

2.8.3   Value of GIS and other GI visualization systems

Many studies have looked at the value of geographic information systems (GISs), i.e., IT tools, models, and decision support systems, that use GI in numerous ways. In fact, far more studies focus on the value of the IT aspects of GI use than on the value of geographic information itself, yet without GI there is no purpose in having GIS. Because valuing a computer-based tool is not the same as valuing the data used by that tool, detailed analysis of the value of GIS as a specialist technology is not covered here. Rather, the reader is directed to review the case studies, comments, or guides offered by numerous organizations from both inside and outside the GIS industry, e.g., the U.S. Federal Chief Information Officers Council (CIO Council, 2004a, 2004b), the Gartner Group (Kreizman, 2001), and myriad stories in dozens of GIS trade magazines over the past decade. A GIS cost–benefit assessment carried out by the Montgomery County Council (1999) in Maryland lists the main benefits of applying GIS in local government, which appear in most similar studies, as:

- Enabling improvements in existing operations
- Adding capabilities that are not available in a non-GIS environment
- Improving the response time and capability to unexpected or emergency situations
- Delivery of intangible improvements to service provision
- Generating revenue through sale of data and products

Similar findings were reported by McInnis and Blundell (1998) for in-depth cost–benefit studies across local and state governments in Montana.

We do recognize that much geographic information is of limited value by itself, especially if one is referring to the location attribute alone. Exceptions are perhaps addresses and raster images that can be viewed by decision makers, which help orientate them with regard to the physical environment. However, it takes software other than only GIS tools to be able to process, store, transmit, and manage GI successfully and efficiently. Thus, one cannot completely ignore the added value for GI that is provided by advances in GIS and related information processing and management technologies, such as spatially aware database management systems and new GI visualization technology and systems intended mainly for laypersons' use, e.g., services such as Google Earth and Microsoft's Virtual Earth, as well as numerous Web map serving tools now available both commercially and for free.

For example, a national thematic data atlas, e.g., for land use, geodemographics, or coastal/marine data, based on standardized Web-based GI visualization tools, permits nonexperts to view spatial data for which access would otherwise be limited to expert users with specialist knowledge of using often complex, proprietary GIS systems. According to O'Dea et al. (2004, p. 6), from their experience with a national marine data atlas, the value of data is increased "when users have the ability to see its potential when displayed with other available atlas layers." The U.K. Environment Agency's What's in Your Back Yard (WIYBY) free online service is another excellent example of wider use of PSI originally collected to aid the agency in its mandated tasks, yet whose value to all citizens is greatly enhanced by user-friendly presentation and easy availability via the WIYBY interface (U.K. Environment Agency, 2007). A further example from the U.K. is the retail census database of town center statistics (Geofutures, 2007) developed by the U.K. Department of Communities and Local Government, which combines employment and retail turnover data from the Office of National Statistics with floor space and rateable value data from the Valuation Office Agency. The U.K. Advisory Panel on Public Sector Information, in its 2006 annual report, notes that "further value was added to this PSI service by making it available online using Google Maps" (APPSI, 2006, p. 32).

On the other hand, improper use of GIS, especially visualization systems that may be used by nonexperts, can devalue both GI and GIS as analytical or decision support tools, by destroying confidence in the underlying data and tools. For example, visualization systems that permit users to view data sets from different sources or of unknown and variable quality can lead to misinterpretation of the data and poor decision making. Also, displaying data collected at different geographic scales as though it were harmonized to one scale can lead to equally embarrassing results, such as houses placed in the center of major roads or coastal hotels appearing to sit hundreds of meters offshore.

2.9   Conclusions

So many different types of information are labeled "geographic" that it is very difficult to assign a value to GI in general terms. Direct monetary value based on GI data sales revenue tells only a very small part of the tale. Such figures are anyway confused since they include fully commercial data sales by private industry and sales of GI by governments at local, regional, and national levels, plus by one government agency to another, raising the issue of multiple accounting of the sale and perceived value of GI if sales figures are accepted as a valid surrogate for value.

Numerous examples exist, across widely varying sectors of industry and economies that create or use GI, that report benefit–cost ratios ranging from 2:1 to 150:1 for GI. Significantly, no study or report identified during the research for this book has found a single negative benefit–cost ratio for GI.

Understanding and, more importantly, accepting a stated value for various types of GI is crucial to being able to perform cost–benefit analyses (CBAs) for systems or services that incorporate such information, as is further explored in Chapter 6. A cost–benefit analysis is only as good as the assumptions that underpin the analysis methodology, and where benefits are concerned, much depends upon the value assigned to the geographic information used in the project or program assessed. Since value is one of the determinants of price and potential charging regimes, as discussed in Chapter 4, it also has an impact on the cost assumptions in a CBA.

Returning to our opening quote from Barr and Masser (1996) that unused information has no value, we next consider that for information to be used, it must first be accessible under terms that permit its exploitation for both the originally intended use and new, innovative purposes. The impact of different access and exploitation policies, especially for public sector GI, is the focus of Chapters 3 and 4, which examine business issues concerning access, demand, cost, and pricing for geographic information, especially those arising from within the public sector.

References

AGI. 1996. *Guidelines for Geographic Information Content and Quality: For Those Who Use, Hold or Need to Acquire Geographic Information.* Association for Geographic Information, London.

Angeletos, G.-M. and A. Pavan. 2006. Efficient use of information and social value of information. *Econometrica*, 75(4), July 2007, 1103–1142 http://faculty.north western.edu/~apa522/AngeletosPavan_Ecma_Final.pdf (accessed September 20, 2007).

Anon. 2003. *McGraw-Hill Dictionary of Scientific and Technical Terms*. McGraw-Hill Companies, New York. http://www.answers.com/topic/value-theory#top (accessed May 10, 2007).

ANZLIC. 2005. *ANZLIC's Vision: Strategic Plan 2005–2010*. Australia New Zealand Land Information Council, 2005. http://www.anzlic.org.au/about_ANZLIC_ strategic2005-2010.html (accessed July 12, 2006).

APPSI. 2005, March 17. *Managing Public Sector Information More Coherently: 2nd Annual Seminar*. Advisory Panel on Public Sector Information, Norwich, U.K. http:// www.appsi.gov.uk/minutes/2005-03-17.pdf (accessed August 15, 2006).

APPSI. 2006. *Realising the Value of Public Sector Information: Annual Report 2006*. Advisory Panel on Public Sector Information, Norwich, U.K. http://www.appsi.gov. uk/reports/annual-report-2006.pdf (accessed February 10, 2007).

Barr, R. and I. Masser. 1996. The economic nature of geographic information: a theoretical perspective. In *Proceedings of the GIS Research U.K. 1996 Conference*. University of Kent, UK, April 1996, pp. 59–66.

Bartlett, D. 1999. Working on the frontiers of science: applying GIS to the coastal zone. In *Marine and Coastal Geographical Systems*, Wright, D. and D. Bartlett (Eds.). Taylor & Francis, London, pp. 11–24.

Bhagwat, S.B. and V.C. Ipe. 2000. *The Economic Benefits of Detailed Geologic Mapping to Kentucky: Illinois State Geological Survey Special Report 3*.

Bryson, J. 2001. Value and performance in the IT society. In *Proceedings of ALIA 2000 Conference, Capitalising on Knowledge: The Information Profession in the 21st Century*, October 24–26, 2000. Australian Library and Information Association. http://conferences.alia.org.au/alia2000/proceedings/jo.bryson1.html (accessed March 15, 2006).

Cambridge Advanced Learner's Dictionary. 2005. Cambridge, U.K.: Cambridge University Press. http://dictionary.cambridge.org/ (accessed April 5, 2007).

CIO Council. 2004a. *The Value of IT Investments: It's Not Just Return on Investment*. Chief Information Officers Council, Best Practice Committee. http://www.cio. gov/documents/TheValueof_IT_Investments.pdf (accessed August 26, 2006).

CIO Council. 2004b. *VMM Value Measuring Methodology How-To Guide*. Chief Information Officers Council, Best Practice Committee. http://www.cio.gov/documents/ValueMeasuring_Methodology_HowToGuide_Oct_2002.pdf (accessed August 26, 2006).

Cobb, J.C. 2002. The value of geologic maps and the need for digitally vectorized data. In *Digital Mapping Techniques '02: Workshop Proceedings of the U.S. Geological Survey*, Open-File Report 02-370. U.S. Geological Survey. http://pubs.usgs. gov/of/2002/of02-370/cobb.html (accessed June 15, 2006).

Cowen, T. 2002. Public goods and externalities. In *The Concise Encyclopaedia of Economics*. http://www.econlib.org/library/ENC/PublicGoodsandExternalities. html (accessed May 10, 2007).

Daratech. 2006. *GIS/Geospatial Markets and Opportunities*. Daratech, Inc., Cambridge, MA. http://www.daratech.com/research/gis/2006,(accessed August 6, 2006).

European Parliament. 2000. *Resolution on the Commission Communication to the Council, the European Parliament, the Economic and Social Committee and the Committee of the Regions on the "Priorities in EU Road Safety: Progress Report and Ranking of Actions,"*

COM(2000) 125-C5-0248/2000 — 2000/2136(COS). http://www.europarl. europa.eu/sides/getDoc.do?pubRef=_//EP//NONSGML+REPORT+A5_2000_ 0381+0+DOC+WORD+V0//EN (accessed September 20, 2007).

FGDC. 2004, May 3. Public Review: Security Concerns about Access to Geospatial Data. Federal Geographic Data Committee. http://www.fgdc.gov/whatsnew/ whatsnew.html (accessed May 11, 2004).

Generalitat Catalunya. 2005. *LAW 16/2005 of 27th December, on Geographical Information and the Cartographic Institute of Catalonia*. President's Office, Generalitat de Catalunya, Barcelona, Spain. http://www.icc.es/web/content/pdf/ca/common/icc/Llei_Info_geografica_ICC_271205.pdf (accessed March 18, 2007).

Geofutures. 2007. DCLG Town Centres Statistics online service. http://www.geofutures.com/online/towncentres.html (accessed March 15, 2007).

Geoscience Australia. 2002, July. *House of Representatives Standing Committee Inquiry into Resources Exploration Impediments*. 49 pages.

Hirschleifer, J. 1971. The private and social value of information and the reward to inventive activity. *American Economic Review*, 61: 561–574.

Joffe, B. and T. Bacastow. 2005. Geodata access while managing digital rights: the Open Data Consortium's reference model. *URISA Journal*. (Submitted for peer review January 10, 2005).

Kreizman, G. 2001. The Value of GIS in Government. http://ontogeo.ntua.gr/nagii/ GIS_in_Government.pdf (accessed August 2, 2006).

Larousse. 1997. *Larousse English Dictionary*. Larousse CD.

Lash, S. 2002. *Critique of Information*. Sage, London.

Longley, P., M. Goodchild, D. Maguire, and D. Rhind. 2001. *Geographic Information Systems and Science*. Wiley & Sons, Chichester, U.K.

McInnis, L. and S. Blundell. 1998. *Analysis of Geographic Information Systems (GIS) Implementations in State and County Governments of Montana*. The Montana Geographic Information Council, Helena, MT. http://www.nysgis.state.ny.us/coordinationprogram/reports/montana/ (accessed July 7, 2006).

Merriam-Webster Dictionary Online. 2007. Merriam Webster, Inc., Springfield, MA. http://www.m-w.com/dictionary/value (accessed February 2, 2007).

MetroGIS. 1999, September 23. MetroGIS Coordinating Committee meeting. http:// www.state.mn.us/intergov/metrogis/teams/cc/meetings/m_09_23_99.shtml (accessed August 12, 2006).

Montgomery County Council, Maryland. 1999, April 12. *Geographic Information System Cost/Benefit Assessment Report M-NCPPC*. Montgomery County Council, Management and Fiscal Policy Committee. http://www.ec-gis.org/sdi// ws/costbenefit2006/reference/gisbenefits_montgomery_county.pdf (accessed August 22, 2006).

NAVTEQ. 2007. The NAVTEQ Database. http://www.navteq.com/about/database_ coverage.html (accessed April 21, 2007).

NetMBA.com. 2007. The Value Chain. http://www.netmba.com/strategy/value-chain/ (accessed August 26, 2006).

NZ Police. 2004. *Annual Report 2004*. New Zealand Police. http://www.police.govt. nz/resources/2004/annual-report/annual_report.pdf (accessed September 15, 2007).

O'Dea, L., V. Cummins, and N. Dwyer. 2004. Developing an Informational Web Portal for Coastal Data in Ireland: Data Issues in the Marine Irish Digital Atlas. Paper presented at Oceanology International 2004, London. http://mida.ucc. ie/assets/documents/oceanology2004_mida.pdf (accessed December 14, 2006).

Oelschlager, F. 2004, October 3–6. Enterprise Information Integration: Enabling the Information Value Chain to Achieve Business Optimization. Paper presented at the Primavera 21st Annual Conference, New Orleans, LA. http://www.primavera.com/partners/files/Enterprise_Information.pdf (accessed August 20, 2006).

OFT. 2006. *The Commercial Use of Public Sector Information (CUPI)*. OFT, London. http://www.oft.gov.uk/advice_and_resources/publications/reports/consumer-protection/oft861 (accessed May 10, 2007).

OMB. 2005. *Enterprise Architecture Assessment Framework*, version 1.5. OMB, FEA Program Management Office, Washington, DC. http://www.whitehouse.gov/omb/egov/documents/OMB_Enterprise_Architecture_Assessment_v1.5_FINAL.pdf (accessed August 12, 2006).

Onsrud, H. 2004. Exploring the Library Metaphor in Developing a More Inclusive NSDI. GeoData Alliance website, http://www.geoall.net/library_harlanonsrud.html (accessed May 13, 2007).

Ordnance Survey. 2007. Welcome to OS MasterMap (online portal). http://www.ordnancesurvey.co.uk/oswebsite/products/osmastermap/ (accessed March 15, 2007).

Oxera. 1999. The Economic Contribution of Ordnance Survey GB, Oxera Economic Research Associated, Ltd., Oxford, UK. http://www.ordnancesurvey.co.uk/aboutus/reports/oxera/oxera.pdf (accessed September 15, 2007).

Phillips, R.L. 2001. The management information value chain. *Perspectives*, 3. http://pages.stern.nyu.edu/~abernste/teaching/Spring2001/MIVC.htm (accessed August 24, 2006)

PIRA. 2000, September 20. Commercial Exploitation of Europe's Public Sector Information, executive summary and final report. PIRA International Ltd. and University of East Anglia for the European Commission, DG Information Society. ftp://ftp.cordis.europa.eu/pub/econtent/docs/commercial_final_report.pdf (accessed Sept. 8, 2007).

Pluijmers, Y. and P. Weiss. 2001. *Borders in Cyberspace: Conflicting Government Information Policies and Their Economic Impacts*. For PricewaterhouseCoopers, National Weather Service, Washington, DC.

Porter, M.E. 1985. *Competitive Advantage*. The Free Press, New York.

RoadPeace. 2003. Cars of the Future: Memorandum by RoadPeace (CAR 34). U.K. House of Commons Select Committee on Transport. http://www.publications.parliament.uk/pa/cm200304/cmselect/cmtran/319/319we36.htm#note69 (accessed January 21, 2007).

Samuelson, P. 1954. *The pure theory of public expenditure. Review of Economics and Statistics, 36:* 387–389.

Shapiro, C. and H. Varian. 1999. *Information Rules*. Harvard Business School Press, Boston.

Spataro, J. and B. Crow. 2002. A framework for understanding the information management market. *The Gilbane Report*, 10. http://gilbane.com/gilbane_report.pl/79/article.html (accessed August 20, 2006).

U.K. Environment Agency. 2007. What's in Your Back Yard? (online data service). http://www.environment-agency.gov.uk/maps/ (accessed February 12, 2007).

U.S. Federal Highway Administration. 1998. *Value of Information and Information Services*, Publication FHWA-SA-99-038. Volpe National Transportation Systems Center, U.S. Department of Transportation, Research and Special Programs Administration, Washington, DC. https://www.fhwa.dot.gov/reports/viis.pdf (accessed August 12, 2006).

Utah Geological Survey. 2006. Why are geologic maps important? In *Geologic Maps: What Are You Standing On?* Utah Geological Survey Public Information Series 66, p. 4. http://geology.utah.gov/online/pi-66/pi66pg4.htm (accessed June 10, 2006).

Van Loenen, B. 2006. *Developing Geographic Information Infrastructures*. Delft University Press, Delft, The Netherlands. http://www.sdic-more.org/docs/van-loenen_phd.pdf (accessed February 12, 2007).

Vowles, G. (Ed.). 2006. *Geospatial Digital Rights Reference Model (GeoDRM RM) (06-004r3)*. Open Geospatial Data Consortium, Inc., Wayland, MA. http://portal.opengeospatial.org/files/?artifact_id=14085 (accessed August 12, 2006).

Wagner, R. 2006. *OWS-3 GeoDRM Thread Activity and Interoperability Program Report: Access Control & Terms of Use (ToU) "Click-Through" IPR Management*, version 1.0.0. Open Geospatial Consortium, Inc., Wayland, MA. http://portal.opengeospatial.org/files/?artifact_id=13958 (accessed August 12, 2006).

Wehn de Montalvo, U., E. van de Kar, and C.F. Maitland. 2004. Resource-based interdependencies in value networks for mobile Internet services. In *ACM International Conference Proceeding Series*, Vol. 60, *Proceedings of the 6th International Conference on Electronic Commerce*. http://portal.acm.org/ft_gateway.cfm?id=1052300&type=pdf&coll=GUIDE&dl=GUIDE&CFID=22551963&CFTOKEN=90347267 (accessed August 22, 2006).

Weiss, P. 2002. *Borders in Cyberspace: Conflicting Public Sector Information Policies and Their Economic Impacts*, summary report. National Weather Service, Washington, DC.

Wikipedia. 2007. http://en.wikipedia.org/wiki/Socioeconomic (accessed April 22, 2007).

Wikipedia. 2007. http://en.wikipedia.org/wiki/Public_goods (accessed April 22, 2007).

chapter three

The business of GI:
No such thing as a free lunch

3.1   The turbulent interplay of price, cost, and value

Let us state something at the outset regarding the next two chapters. We are not against freely available data. We are not against a free lunch. We do not hold any particular doctrine about whether geographic information collected in the public sector should be freely available or available through a commercial cost or any cost level in between, i.e., cost of reproduction and dissemination, etc. We believe in freedom of information, but do not necessarily assume that the information always should, or needs to be, made available free of any charge. In all information collection and dissemination transactions there are costs, and someone, somewhere, has to pay for them. Admittedly, the emergence of information technologies and electronic networks has reduced some of the costs dramatically, and as we see with e-commerce and the media, user consumption patterns have changed, as have users' willingness to pay charges. This chapter, then, is a no-holds-barred exploration, but please do not take it personally. What we hope to achieve is to set the scene for a reasoned, objective debate within the widest range of geographic information (GI) stakeholders as possible, whether in government, business, or civil society, whether as owners, users, or custodians.

The impact of the Internet on the pricing of information and communication has been substantial. We can now access information that previously was the expensive and protected domain of specialists, for example, looking online at flight tracking at major airports (Floweb, 2006). Built on the emerging Google Maps and Google Earth (Google, 2006) innovations, Floweb continues a process where the price of information and the quality and availability of information bring previously premium products and applications into the mass market. Computer flight simulators and in-car navigation are two examples of technologies that have experienced significant cost reduction. They previously were expensive, premium technologies. Automobiles have been demonstrating this trend for years, with air-conditioning and antilock brakes, which were previously available only on high-price executive cars, in the context of the innovation curve, but which are now normal fittings.

Floweb also continues a process whereby the uncertain and unwelcome aspects of globalization, such as global terrorism, present ethical and political challenges to governments, particularly where readily available information may assist terrorism and crime. We reviewed those processes following the events of September 11, 2001 (Blakemore and Longhorn, 2001). In September 2005, the government of South Korea was upset because Google Earth showed the locations of sensitive military installations (Haines, 2005). In 2006, Google Earth was used to detect a Chinese military model of disputed territory on the border with India (Haines, 2006). The U.S. government has reserved the right to shut down the GPS satellites at a time of national emergency (Wired, 2004), a fear that in part had already motivated Europe to launch its own navigational satellite system, Galileo (Shachtman, 2004). The U.S. government also started to remove information from the public domain that was deemed to be supportive to the planning of terrorism (FGDC, 2004a).

We have become used to reading newspapers online free of charge. An information paradox has developed whereby we often are still willing to pay real money to receive a newspaper delivered to our residence, whereas we can read the information online, often at no cost, well before the relatively outdated newspaper has arrived. Such is the disruptive pace of change that there are some fears that wikis, blogs, and citizen journalism may kill off the newspaper in its traditional form, for how will newspapers be able to obtain the revenue to invest in their production if online access is free? Far from killing off newspapers as a genre, however, the *Economist* argues that "for hard-news reporting — as opposed to comment — the results of net journalism have admittedly been limited" (Economist, 2006e). In effect, the *Economist* is arguing that quality, continuity, and robustness will continue to have a significant market demand.

A similar finding was reported by Michael Blakemore and Sinclair Sutherland (2005), in the context of their experiences running the U.K. online labor market statistics service NOMIS. When, in 2000, U.K. National Statistics made the service free of charge, the expectation was that the removal of charging would lead to an explosion of usage. However, while the number of users did increase, the actual usage did not increase proportionately. Much usage was one-off, and the users who previously had paid the most for high levels of usage now had diminished power in influencing service development; whereas their feedback had been significant before 2000 in maintaining quality control and prioritizing service developments.

While many free-GI proponents defend their stance on the premise that more information, made available free of charge, will lead to more usage and societal impact, we do not infer that there is an automatic, direct, and immutable link between free-of-cost (to the end user) access to GI and increased usage or societal impact. Consider U.K. public museums, for example. Under the Thatcher government, with its mantra resembling "If you need it, pay for it; if you cannot pay for it, you do not really need it," charges were introduced for entry to museums where there had previously been no charge. Not surprisingly, entry levels dramatically reduced, and in 2001 the New Labour Government of Tony Blair abolished the charges. A report 5 years after access was again made free indicated that there was an 83% increase in visits, some 30 million extra visits over 5 years (Brown, 2006). So far, so good. Lower prices often lead to more consumption. However, while U.K. Culture Secretary Tessa Jowell argued that these were "inspirational figures ... there is a real appetite for serious culture in this country," (Brown, 2006) there was no clear evidence whether the figures represented more visits by the same people, i.e., those who had been willing to pay in the past, and therefore cost of free entry was subsidized for this more affluent segment of society, or whether the visits by people who had been previously excluded had resulted in a cultural impact. In effect, did the measurable transactions of people through the museum door translate into societal value? Furthermore, the museum income streams were now largely dependent on a central government grant, plus income from areas such as merchandising and special exhibitions — not everything was free and some premium facilities were made available only at a cost. As a result, there was now concern that the costs of meeting the increased demand were not being met, and that the government was considering cutting the central grant in 2007, leading to the possibility that charges would be reintroduced. As should be recognized by everyone, government taxation coffers are not limitless, and demands upon the government purse are many and varied. These demands are often fulfilled using cost–benefit considerations characterized by multiple interpretations, from the purely financial, e.g., 10 million euro spent on transport today will generate 100 million euro economic benefit overall, to the more subjective and emotive, e.g., 10 million euro spent today on pay for more doctors, nurses, or health care will prevent a statistically calculable number of citizens' deaths.

It is the turbulent interaction of supply, demand, and resource, combined with the almost religious zeal of policy positions (charge a fee or make it free) that we investigate in this chapter. We introduced theories of economic value of information earlier in this book, and here we relate the theory to the operational practice of politics, business, and money. For example, in 2006 the U.K. Office of Fair Trading (OFT) investigated the relative success of commodified data availability in the U.K. by public sector information holders (PSIHs) and found that more competition in data provision, not necessarily for free but at justifiable costs, such as cost of dissemination, "could benefit the UK economy by around £1 billion a year" (OFT, 2006). The restricting factors were more in areas of anticompetitive behavior by information owners who needed to maximize prices and protect market position so that they could meet government income targets, the principle under which U.K. government trading funds operate. The OFT report implies that it is when charging is applied in this context that data access diminishes, with detrimental effect on the economy. However, the interpretation by those who promote free access to data, such as the Free Our Data campaign in the UK, is very clear: "public bodies are secretive about the data they hold, restrictive in the way they license it, and may be abusing their position as monopolies" (Cross, 2006).

Price and value interplay in complex ways in the information society. Something that is free may have high value, and not necessarily vice versa, and something that has low value can generate much higher value. In 2006, one person sold a single paper clip and purchased a house in the town of Kipling, Saskatchewan, Canada (BBC, 2006f). Admittedly this was not a direct purchase, but a series of trades that in truth did not have direct value relationships. The first online trade was the paper clip for a novelty pen, and the 14th and final trade was a role in a Hollywood movie for the house. The cost–price–value interplay involved many processes. The fact that the initiative gained significant media attention encouraged people to make trades, to reap the value of 5 minutes of fame. As the trades progressed, the value exchange became more significant, driven perhaps by the trading of intangibles, an experience rather than an object that may not have been directly purchased by the owners, for example, an afternoon in the company of the rock star Alice Cooper or the value of temporary fame in the film role.

After decades of having to pay for telephone communications, either by time or subscription, will Voice over Internet Protocol (VoIP), as championed by the Skype service, "herald the slow death of traditional telephony" (Economist, 2005a)? Skype, however, was never truly free, but was just not exacting a direct charge to most users. Those who use Skype are in effect donating some of their resources to the service, which as a result has almost no marginal costs when expanding the service, because "users 'bring' their own computers and internet connections or marketing (users invite each other)" (Economist, 2005b). Skype uses your computer resources as part of its virtual infrastructure, avoiding the significant infrastructure investment costs. That is laudable, and conceptually Skype is a business version of the much lauded SETI (Search for Extra-Terrestrial Intelligence) project. This is an example of *gifting technology*, where people donate spare resources on their PCs to allow the SETI project to process huge amounts of data in search of extraterrestrial intelligence (McGee and Skågeby, 2004). Another gifting technology project is Climateprediction*, which also uses the computing capacity gifted by individuals (BBC, 2002).

Problems have occurred, however, when many people use Skype at work, and the resource impact can be significant — each user in effect is donating a proportion of the corporate network to Skype (Crampton, 2006). Business strategy also has an impact in pricing, for Skype was purchased in October 2005 by eBay, and the purchase price of $2.5 billion needed to be recouped somehow: an income stream is a classic mechanism. Therefore, from the start of 2007, calls made to landlines in the U.S. and Canada are no longer free, but are charged at a flat fee of $30 a year, being "part of a broader strategy by eBay to expand Skype's product offerings and revenue" (Richtel, 2006). The flat fee, and the level of it, is an elegant mediation between consumer resistance to the introduction of fees. It is not so high as to deter the majority of

---

* http://www.climateprediction.net/.

users, and efficiency for the business, and it is a single transaction to process, and the volume of payment transactions should generate significant levels of income for the business to invest into infrastructure. Skype thus provides a good example of the key theme of this chapter: the lunch is seldom free — it is just paid for in different ways.

The death of a genre, when examined historically, is more a case of a disruptive technology threatening the existing status quo. This leads to a nervous and often defensive reaction by those with vested interests, thus resulting in a mutation of the technology to provide greater market access — newspapers, television, and telephones all have followed such a path. The equivalent process seen in geographical information is the expectation that data will be available at increasingly low cost, or even free of charge. Therefore, this chapter aims to build a conceptual framework to explain the emotive, often polarized debate about whether public sector information (PSI) — of which government GI (PSGI) is a component, and we shall use these two acronyms and the terms *data* and *information* interchangeably — should be freely available to citizens and businesses. The debate is often complicated by lack of prior definition of the term *free* used by those deliberating different issues, such as *freely available, free of charge, free of restrictions on use, free of restrictions on reuse* (exploitation), and *readily available* — the last term implying that the data may be free of charge, but not available quickly enough or in appropriate formats for use or reuse.

3.2 Access, demand, resource, and information supply

At the outset we hypothesize that providing access to information is an economic and political contest between resource allocation and user demand, as already indicated in the few cases presented in the previous section. The overall perspective will be one of realism. While many cost–benefit arguments have been proposed for making information freely available (see Chapter 6), thus generating significant use of GI, there is a real difficulty in then ensuring that information is both up to date and targeted to the broad set of user needs, let alone those needs that are of most value to society as a whole. The contest is nowhere more evident than in core government services such as public health. National health services have perversely been focused on both public health, through processes such as immunization, and illness, i.e., treating people when they are unwell. These are often services that are primarily centrally funded through taxation and which promote themselves as being largely free at the point of demand. The result is, inevitably, a mismatch between supply and demand, both structurally and spatially. Attempts to diminish the mismatch include:

- Administrative reform, e.g., creating centralized health trusts in the U.K. system to supposedly reduce administrative cost

- Contracting out some service provision, e.g., paying private health companies, or even health centers and hospitals abroad, to treat U.K. patients unable to be serviced by the national health system
- Technology use, a double-edged sword, since it can both save costs and impose new ones through advanced and expensive technologies and drugs
- Manipulating waiting list rules or statistics

Where these strategies have little impact is on the behaviors of the users. This can generate superficial debates about whether we should stop treating smoking or alcohol-related diseases because they are self-inflicted. The rebuttal is that so are sports-related injuries. The mismatch is exacerbated further by other lifestyle issues, such as diet. In the U.K., the cost of treating obesity consumed 9% of the National Health Service (NHS) budget in 2005 and "could bankrupt the NHS if left unchecked" (BBC, 2006h). With these huge dilemmas facing them, it is therefore not surprising that governments may argue that charges by the national mapping service, the Ordnance Survey of Great Britain (OSGB), are trivial, since OSGB costs a bit over £100 million a year to run compared to the NHS cost of £76.4 billion. In the current political and financial climate, concerns about information charges for PSGI of around 0.13% of NHS costs really do not register on the policy horizon.

On one side of the information contest the data producers have a budget to collect, structure, and sometimes disseminate information. On the other side of the contest are those people and organizations that wish to use information and therefore place demands on the producers. The demands may simply be that they want to use the data, in which case the data may be available at minimal (but not zero) distribution cost via an Internet site. As discussed in Chapter 2, the process of disseminating data incurs what theorist Scott Lash calls exchange value (Lash, 2002). Once the data are used, the results of the use generate added value, which Lash calls use value. For example, a data set of road lines and names can be sold at one price, but when the data are embedded in a vehicle navigation device, the value of the data is higher. The exchange value of historical information or information already legally in the public domain may be zero, e.g., where no copyright implications exist, so little or no acquisition cost is incurred. However, realizing the use value of the information incurs sunk costs of database preparation and maintenance, plus access and distribution costs, which most probably generates valuable use to someone; otherwise, the service or product would not be created in the first place.

Hence, the public availability of the 1871 Census of Population (BBC, 2005b) or the *Domesday Book* of 1086 (Archives, 2006) in the U.K. are semicommercial services where basic information is free, but full detail is available for a charge, where the charge is for providing the information in a usable format. However, this charging model may be destabilized if Google proceeds to digitize large volumes of historical material (Roush, 2005a). Google's intention is to scan millions of books, providing access to the full text for those out of copyright and extracts from those under copyright (BBC, 2006d), via its Books Project, with university partners such as Oxford, Harvard, Stanford, the University of Michigan, and the University of California, as well as the New York Public Library.

For there to be a reason to engage in information exchange then, one expects that use value of the information should be higher than the exchange value, yet use value is "highly dispersed and difficult to trace" (Lash, 2002). Lash notes the benefits to an economy through more use value, e.g., more business, more employment, more tax income perhaps, but also that the highly distributed nature of use value places new and increasing demands on the data suppliers, e.g., the needs of a growing range of application areas such as mobile navigation or geosurveillance. For example, users may want to receive advice, or they may want to suggest changes to the data and improvements in quality. That leads to the basic question: How can the demands of use value be resourced by data suppliers? This is at the core of the debate.

The contest can be distorted in either direction by either player, producer or user. It is easy to inflate demand for information either by offering new services to new users of data, a positive development, or through permitting or encouraging mendacious requests for data that impose onerous demands on data suppliers, a negative development. The availability of information, even when available through freedom of information (FOI) legislation, can be suppressed by changing the rules of access, reducing the finance available to enable the dissemination of information, discontinuing a data series, or reclassifying information to fall within the various exceptions existent in most FOI legislation. For example, in June 2006 a citizen request in the village of Lakemoor, IL, was charged at 17 U.S. cents per page (Klapperich, 2006). The reporter investigating the case found that even the commercial copy shops in the area charged a maximum of 8 U.S. cents, and another citizen was provided with the costs that Lakemoor budgets for copying, which was 1 U.S. cent per page. Superficially, then, the local government was profiting under FOI.

Mendacious requests work the other way, demanding unacceptable amounts of time. In June 2006, the information commissioner for Scotland ruled on a case in which a citizen had requested 13 items of information about all the property in the Tayside valuation area (Dunion, 2006) — a significant amount of information. The financial threshold, calculated by staff time and administrative costs in complying with the request, beyond which a request can be refused, is £600 under U.K. FOI legislation, and the actual calculation of costs to comply with the request was £898.08. The request was refused, and the applicant appealed, leading to this judgment. So, legislation that is intended to liberate data was then leading to a long dispute over £298.08 beyond the threshold, involving a local government assessor and the Scottish information commissioner. The 2004 annual accounts for the

Scottish information commissioner* indicate that he was paid a salary of about £75,000 plus performance bonuses, which works out at about £340 a day (220 working days a year). Add an hour of his time, plus all the other staff time taken in assessing and challenging the request and complaint, and the cost of arguing over £298.08 was probably more than 10 times that amount. Still, we must have rules, must we not, even where the cost of defending an arbitrary rule is a significant cost to the taxpayer?

On the other hand, criteria can be adjusted in favor of government, as was the case in the U.K. during 2006 with a proposal to charge a flat fee for all FOI requests, which, given experience in Ireland, would lead to requests dropping by 30% (Cracknell, 2006). In October 2006, a review of FOI costing rules by the U.K. government was announced (DCA, 2006), but it was difficult to see how the demand and supply arguments could be mediated when there was an imposed assumption that the average hourly cost for a civil servant to process a request was £254 an hour, and that he or she takes an average of 7.5 hours to process a request (Kablenet, 2006b). If the processing service was put out to commercial tender, would costs be lower?

3.3   Is there such a thing as an informational free lunch: the commons?

The focus of this chapter is on charging for information in the broadest sense. We can build on the examples presented so far regarding the absence of free lunches for most information provision by developing a second hypothesis, i.e., there is no such thing as free PSI, since all PSI is paid for somehow — hence the deliberately provocative title of the chapter.

Claudio Ciborra thought about the pricing of public goods when he asked, "Who should pay for the positive and/or negative externalities created by use?" (Ciborra, 2002, p. 60). He went on to ask how could the "installed base," of existing data production and availability, respond flexibly to the demands for change. Interestingly, Ciborra was very aware that the debates surrounding information are influenced by both rational argument (for example, studies that aim to develop pricing theory or evaluate the economic contribution of data to society — see Chapter 6) and principled positions of belief, which are deeply held beliefs that, for example, democracy is served by making all government data available to citizens. The principled positions are what Vincent Mosco calls myths, and he is careful to note that myths are not fictional or irrational stories, but like the myths in ancient Greece, they provide an important nexus around which people can gather, discuss, and construct beliefs. Indeed, as Mosco states, "Myths are not true or false, but are dead or alive" (Mosco, 2004, p. 29), and the key question, therefore, is: What keeps myths alive?

---

* http://www.itspublicknowledge.info/Documents/AnnualAccounts04-05.pdf.

One myth already mentioned in this book, and which we will confront again later, is that PSI that is both freely available and free of charge is good for society and the economy. The myth is deeply grounded on U.S. policy, at the federal level, where federal government data (PSI) is available free of charge under the Freedom of Information Act (Congress, 1974), without any copyright restrictions — and hence no restrictions on full exploitation and reuse by others. The Office for Management and Budget (OMB) Circular A-130 to federal agencies states quite clearly that information is a resource that should be available nationally, and that the policy was underpinned by a central assumption that the costs of making the data available would be more than recovered through the benefits that accrued to the nation from data usage (OMB, 1992). There is a powerful logic in the argument, backed up by the statement that the taxpayer has already paid for the collection of the data, and so should not have to pay again to use it.

The free availability of information is an attractive proposition. We can sit in our home offices in Durham (U.K.) and Bredene (Belgium), download U.S. Census* data for 2000, including some very interesting anonymized micro-data files, and set up our own business distributing online value-added reports and services. Granted, we are unlikely to be very successful with that business because there are so many businesses within the U.S. who already market Census products and services. The same example would apply to many potential services built on the back of freely available, current, large-scale data relating to various types of boundaries, real estate transactions, environmental conditions, etc., which are freely available from many of the local and state governments throughout the U.S. under local or state-wide FOI legislation. Since our service could be offered to users — paying customers — via the Internet, we need not be resident in the U.S. to enact some reasonably interesting and potentially lucrative business. The main point is that the U.S. taxpayer has paid for the running of the U.S. Census Bureau, for the collection of the 2000 Census of Population, and we can use the data without contributing anything back to the U.S. Treasury or taxpayers, and similarly for the local and state taxpayers. The services mentioned in the two examples above would not continue to exist unless they provided some use value (mainly to U.S. residents), represented, at a minimum, by some purchase price users were willing to pay for the service (income to us) that is greater than the exchange value (cost to us) for creating the services. By tapping into a much wider, global pool of creative and innovative information market talent and financial resources, does it really matter where the new information service was developed or by whom?

Now we start to build counterarguments in rebuttal. You may reply that it does not matter that we use the data without paying anything, because the cost of getting the data to us is almost zero, using the friction-free dissemination conduit of the Internet. Furthermore, one of the other underlying

---

* http://www.census.gov/main/www/cen2000.html.

assumptions of free data is that it engenders greater democratic participation of citizens because they can more effectively evaluate the performance of their government, and the greater availability of data is positive for educational attainment.

This may be a great idea, but how do we reconcile that view with the fact that at the local level, the level at which participation and governance are usually more evident, the U.S., with all its free data, only managed 38% voter turnout in 1994, whereas the U.K., where chargeable access to much PSI is the norm, managed 69% in 1997*? Why, when all the free federal GI has been available to stimulate democracy over the years, has there been a steady decline in U.S. voter turnout at presidential elections between 1960 and 1990,** with the major participation recovery being after the events of 9/11? Perhaps war and terrorism are a greater motivator for citizen participation than is the ready supply of data? Another argument proposes that all the data help to stimulate economic activity. Maybe, but the economic activity is not generating very equitable benefits, where the "top 1% of Americans now receive about 15% of all income, up from about 8% in the 1960s and 1970s" (Economist, 2006a). How do we relate expected social benefits with reports in the U.S. of "37 million people living in poverty in 2004, or 12.7% of the population," and these numbers continue to increase (BBC, 2005c). Or perhaps voter turnout is simply not a valid proxy for the value to a society of free access to PSI, regardless of the level of government concerned, anymore than is distribution of wealth? Then what success criteria should we be using, and do these vary across different societies and cultures? These are all questions that need addressing in the debate.

Anyway, you say, the added cost for someone to access the U.S. data from the U.K. is so tiny that it does not matter. It does matter, however, when we send e-mails to the nice people at the Census Bureau, or phone them to discuss technical issues related to the data.*** At that point, we are starting to impose a cost on the U.S. taxpayer, who may be waiting in a call queue while we "foreign" non-U.S.-tax-paying freeloaders talk to a specialist, benefiting from increasingly lower telephone call costs, or utilize U.S. government officials' time with e-mails asking for advice. Well, you may rebut, the overall costs for such inquiries may not be large in the overall context of demands on staff time from U.S. citizens and, in fact, probably are not. Furthermore, you may counter, the costs of our requests are more than offset from the broader societal cost benefits of having data freely available, but we are already very skeptical of the social benefit argument given the trends noted above.

---

* http://www.fairvote.org/turnout/compare2.htm and http://www.idea.int/vt/graph_view.cfm?CountryCode=US.

** http://www.ncoc.net/conferences/2004annual.htm.

***Very helpful lists at http://www.census.gov/contacts/www/c-census2000.html and http://www.census.gov/main/www/contacts.html.

Look, you now say, stop picking holes in the broader argument. The U.S. may have issues with poverty levels or distribution of wealth, or with low levels of educational attainment, but what has this to do with data access? As to education, in spite of all this rich GI data and technology, you certainly do have a problem. The National Geographic Roper Survey of geographic literacy in 2006 identifies the lack of a direct link between free data and educational attainment. The survey found that only 37% of young Americans can locate Iraq on a map, in spite of the huge coverage of the war in the media. It also reports that only half of young Americans can locate New York on a map. The report's conclusions were bleak, arguing that the next generation of U.S. business people are unprepared for the global economy "or understanding the relationships among people and places that provide critical context for world events" (GfKRoper, 2006, p. 7).

Now, stop picking on the U.S., you say. Why, we rebut, since in our direct experience over the past decade, the U.S. is held up by commentators globally as a paragon of information availability? What is more, the U.S. is promoting its model widely throughout the world in the context of spatial data infrastructures via the Federal Geographic Data Committee (FGDC), which maintains "an International Activities Coordination staff position to assure continued focus and US leadership presence in global SDI activities" (Schaefer and Moeller, 2000, p. 1). In any case, the very vague economic cost–benefits do not add up when the U.S. economy has experienced uneven development, when the public debt is growing,* and, more importantly, in the context of this debate, it was accepted that much of the freely available and free federal GI was not fit for purpose, e.g., "the average age of the primary topographic series maps is 23 years" (USGS, 2001, pp. 8–9), and "topo maps lagged further and further behind the landscape they represented. Today, the maps are only sporadically updated, and some are 57 years old" (Brown, 2002, p. 1874).

Outdated maps, with no clear investment income stream, presented a bleak position for national mapping. In 2003, this led to a proposal for a form of virtual national map that would be woven together — Weaving a National Map (NRC, 2003) — from other sources. On the one hand, this was an implicit admission that the market had moved away from the U.S. Geological Survey (USGS) to build its own products. On the other hand, this confronted USGS with the fact that it produced topographic data at scales that were of little use at the local level; i.e., 1:24,000 is the most detailed USGS series with national coverage, whereas 1:1,000 to 1:5,000 or larger scale is needed for most local planning, public works monitoring, utilities maintenance, etc. The outcome of this has been a bricolage of large-scale geographic information in the U.S., comprising an uneven coverage of data collected by organizations such as local government, private companies, cities, and utilities. The 2003 report aimed to build on national self-interest, which encouraged these data owners

---

* http://www.brillig.com/debt_clock/.

to allow their data to be used so that USGS could coordinate the production of a national map.

In itself, this act was a further implicit acceptance that the U.S. federal government did not have the funds to invest in its own updating process for the USGS maps, and that the USGS did not have the organizational capacity to produce data quickly enough. Barb Ryan, initial head of the USGS National Map project team, quoted in a 2002 article in *Science* (Brown, 2002, p. 1874), estimated that "delivering the fullscale National Map in 10 years would require $150 million a year — roughly twice the current budget" (2002–2003 annual budget). Within the USGS, the FGDC is tasked with the coordination activities regarding National Spatial Data Infrastructure (NSDI), offering some funding for what they call cooperative partnerships (FGDC, 2004b, 2006) deemed necessary to help data owners with the task of preparing data to National Map metadata and data standards. The federal government is also considering downsizing and outsourcing some of the production functions of USGS (Sternstein, 2005), in a process reminiscent of the U.K. government's downsizing of organizations such as the Ordnance Survey GB (OSGB). Over recent years, OSGB has developed a more market-oriented focus, charging for data use through licensing, agreeing on commercial partnerships with those who are value adding to OS data, and providing the U.K. government with clear value for money and a return on the taxpayers' investment (ODPM, 2004; Survey, 2001). In the U.S., by contrast, there has been strong political opposition even to the closure of one mapping center with 130 employees, and U.S. federal mapping remains imprisoned strategically between inadequate data and resistance to organizational change (Sternstein, 2006b). A further ideological position to change exists with those supporting freedom of information and the free commons, with a person in the U.S. "capturing" "56,000 digital topographic maps (that) have been scattered among many Web sites" and transferring the federal maps to the Internet archive "for free download forever" (Sternstein, 2006a). This may be a fine piece of ideological, community-spirited GI preservation action, but it is difficult to judge the real end-user benefit to be gained from an archive of decaying maps.

There are no 23-year-old data layers in the OSGB database — this high-tech, object-oriented, large-scale database is updated in real time (Survey, 2006b) 50,000 times per day on average. We are not implying that a fully updated database can only be achieved by directly charging for data use. It is more an issue of how an income stream necessary to provide investment in maintenance, enhancement, and updating, plus enrichment of the data set to satisfy evolving new user requests and innovative applications can be achieved. In an ideal world, a government would allocate the necessary funding through taxation. However, most governments are today trying to balance volatile tax flows resulting from fewer people in a workforce, producing less direct taxation, with increasing demands on finance for health, pensions, general social services, environment, homeland security, and sometimes, for some governments, the odd foreign war thrown in for good measure. We generally find that within the economic pressures of globalization, governments are willing at best to fund cheaper free lunches.

As James Carroll wrote, following the debacle surrounding Hurricane Katrina in August 2005, "the United States, after a generation of tax-cutting and downsizing, has eviscerated the public sector's capacity for supporting the common good" (Carroll, 2005). For example, the flood protection infrastructure originally planned for the Lake Pontchartrain and Vicinity Hurricane Protection Project by the Army Corps of Engineers in 1965 was to cost $85 million, to be completed in 13 years. By 1982, 4 years after the initially proposed completion date, the projected cost had risen to $757 million, later reduced to $738 million in 2005, now with a projected completion date (post-Katrina's damage) of 2015. Of this, $458 million had been spent by 2005, yet federal government appropriations had

> generally declined from about $15–20 million annually in the earlier years to about $5–7 million in the last three fiscal years.... The Corps' project fact sheet from May 2005 noted that the President's budget request for fiscal years 2005 and 2006, and the appropriated amount for fiscal year 2005, were insufficient to fund new construction contracts. The Corps had also stated that it could spend $20 million in fiscal year 2006 on the project if the funds were available. The Corps noted that several levees had settled and needed to be raised to provide the level of protection intended by the design. (GAO, 2005)

Yes, hindsight is wonderful, and we do not wish to intrude on the misfortunes of those who suffered death and destruction as a result of Katrina. However, the example demonstrates all too clearly that (1) the true size of large infrastructure project budgets are open to question from the outset and, (2) when push came to shove, funding was reduced at what could have been an important time for the project to be successfully completed. So who gets the free lunch? The Army Corps of Engineers for levee construction that could save thousands of lives in another Katrina, or USGS for 1:25,000-scale topographic data collection?

A hybrid approach, to partial free lunch and partial charged lunch, is seen in the Canadian geographic information infrastructure, developed by Geo-Connections. Two mechanisms are used to develop the infrastructure. First, a central subsidy can be granted where there is a mutual benefit to be gained when another organization develops or deposits data. Second, "GeoConnections agrees to pay for a product or service supplied by the second party," acknowledging (as does the U.S. National Map process) that there is no real commercial benefit for a data producer to deposit data into the infrastructure without some financial incentive (GeoConnections, 2006, p. 24). Yet, again, the free lunch — the eventual provision of an infrastructure for the widest possible benefit to Canadian society and the economy — is being clearly resourced. Similar financial commitments toward construction of SDIs have been demonstrated by governments in the Netherlands at the national level and Catalunya, Spain, at the regional level.

3.4   Resourcing the interfaces between supply, demand, and update

Whatever the approach — direct investment or cooperative agreements — the time horizon for completing a U.S. national map stretches into the distance, and for a long time it will be a Swiss cheese of data domains. The OMB assessment of the National Geological Map program in 2005 noted that only "53% of the United States has geologic map information available needed by customers/decision makers to make land use and water management decisions" (OMB, 2005).

Meanwhile, the U.S. PSI landscape is far more turbulent and complex than before. First, at the federal level, there are budget cuts, increasingly sophisticated and demanding markets for data usage, and collaborative funding strategies. Second, and more importantly, the PSI data held below the federal level are not subject to the free availability legislation, which applies only to federal data, and data selling (commodification) is active in many areas, e.g., the case of San Francisco is provided later in this chapter.

At the federal level, the U.S. Bureau of the Census (USBC), with its decennial Census of Population (the most recent was in 2000), navigates a delicate balance between the costs of ensuring that the Census is enumerated as fully as possible, and allocating its finite budget to priority activities. For example, in 2000, PriceWaterhouseCoopers estimated that if the 2000 Census suffered the same undercount problem as the 1990 Census, then state and local governments would lose $11 billion in federal funding (PricewaterhouseCoopers, 2000). So, should the USBC request extra money to fund better data collection, using a straight cost–benefit argument that $x of investment will generate $x*n in overall benefits to society? It is simple: if the official estimate of your population is lower because of collection error, then you receive less funding where the funding criteria are based on per capita population. Ensuring better data collection inevitably requires more resource, and the full cycle, the total cost of the Census planning collection, and processing "per housing unit of the 2000 census was $56 compared to $32 per housing unit for the 1990 census" (GAO, 2001, p. 2). However, this is potentially perverse, since it is the federal government that pays the funding anyway, so maybe collecting poor data will save money?

A possible hybrid financing model involves partnering with a private sector company that can identify cost–benefits through its investment in a product. At the very least, some form of competitive tendering should help

ensure that the best value for money is obtained for any taxpayer investment. In 2004, the Government Accountability Office (formerly the General
Accounting Office — an interesting refocusing of purpose and title in its own
right) reported on USCB planning for the 2010 Census, requiring a focus on
increasing data relevance, timeliness, coverage, and accuracy, while reducing operational risks (GAO, 2004, p. ii). This has already involved contracting
out the maintenance and development of a key data domain, called TIGER*
(Topologically Integrated Geographic Encoding and Referencing system), to
the Harris Corporation (Harris, 2002, 2006).

What the USCB examples show is that there is not a direct relationship
between central government funding and poor data. The USGS mapping
example must not, therefore, be taken as indicating a general rule that a reliance on government funding produces bad or incomplete data. Nor must the
excellent data produced by OSGB be taken as indicating a general rule that
commodification and commercialization are necessary to produce excellent
data. At this level of argument, the underlying theory, if we can call it that,
is more like political dogma — the U.S. maintains the myth that free data
are essential for society vs. the U.K. government myth that it is good for
you to pay for something you use. The U.K. situation can lead to the government information business approach that characterizes the OSGB, the
Hydrographic Office (including joint ventures like Seazone Solutions Ltd.**),
and the Meteorological Service, which were all considered at the end of 2006
for possible full privatization by the chancellor of the exchequer, subject to
three considerations (Treasury, 2006, p. 146). First, would they still meet public service objectives? Second, can operational efficiencies be achieved if they
are run within the private sector? Third, will they generate finance that can
be reinvested back into core public services? Since, as we will detail below,
even government users pay for access to OSGB data, the issue of whether
the money goes to the government or a private sector company seems not
too problematical. However, that also brings in a useful potential defense
strategy for retaining public ownership of data — the cost of introducing
charging could be seen as adding unnecessary administrative burden. What
actually happens, as seen with the experience of OSGB, is that the strategy is
not linear, but is uneven and often event-led by changing government policy
priorities.

While governments may maintain their myths, they can reinterpret how
their myths are to be performed. For example, continuing the ready supply
of free data in the U.S. has been subject to contest. In 2005, the Republican
senator for Pennsylvania, Rick Santorum, apparently threatened to remove
some weather information from the public domain (Congress, 2005). The
basic reason for the proposal was technological function creep. In the past,
the National Weather Service (NWS) distributed its basic raw information

---

* http://www.census.gov/geo/www/tiger/.
** http://www.seazone.com/.

free of charge, and commercial companies built products on the data — a nice little earner, since the companies paid no money for the data and repaid no income to the NWS. Then, as the price of IT came down and functionality increased, "advances in computer graphics and software have enabled the Weather Service to easily package its information in a more appealing way" (Withers, 2005). In other words, it became more possible for the NWS to offer data analysis, in the form of weather forecasts that the public could understand, at a much reduced cost than before, especially via the Web — so why not provide this as a public service? The progressive creep of the NWS into product development was then called foul by industry, claiming unfair competition by the taxpayer-funded NWS. So the Santorum proposal takes us back to the position that if it is free, then just let the basic data out free, and do not develop value-added product lines — that is the role of business.

3.5 *Can a free lunch be sustained?*

The previous section entailed a long discussion, and while it may seem to be hostile to the U.S. position (please note that it is not meant to be), the rationale for making these points is to set the scene for a deconstruction of the myth and an exploration of price and cost of information in a turbulent globalized marketplace. We will now discuss examples of free data. After all, we are accustomed to increasingly rich information sources free of charge on the Internet. They may be free, but for how long? The experience of Wikipedia will be one case study where something free, and openly democratic, became so large that it needed to start formalizing its activities in 2006. Wikipedia was built on the free-of-charge investment by those who wrote the entries, and was then available free via the Internet. That worked in a satisfactory way, but as the content expanded, there was not a commensurate increase in management resource to ensure quality control — not surprising since without an income stream there is nothing to fund management, and the "brand" of Wikipedia has to be maintained on an assumption of vested and ethical self-interest.

In 2006, an outbreak of deliberately distorted entries, and the deliberate injection of incorrect information (Martin, 2006), forced Wikipedia to become much more structured in its editorial policy. Putting these developments into overall context of informational trust and reliability, Lee Shaker concluded that "though developing technologies like blogs and wikis have great promise, they also are nascent and unreliable at this point" (Shaker, 2006). The rapid, and uncertain, emergence of threats to the free, though trusted, Wikipedia brand forced a strategic rethink by the "owners." By August 2006, Wikipedia had ceased to be the anarchic "anyone can contribute" brand; a much more conventional approach was emerging where "a cadre of privileged users will supervise what appears" (Thompson, 2006b).

Many Internet free services are underpinned by both very low cost IT and increasingly low cost labor. Wikipedia used no-cost labor to create content, and the no-cost labor had no intellectual property rights to the content either. In its less troubled days, Wikipedia genuinely represented the goal of an information commons (Onsrud, 1998). Look at call centers, for example, which migrated from North America and Europe to India, but then started to migrate from India to even lower cost locations, a process that may continue until, as the *Economist* noted, we will eventually all work for free. The owner of several Bangalore call centers, faced with the possibility of these moving to Indonesia, e.g., as itinerant businesses follow the cheaper labor market option, said "it's hard to know where it will all end. Is there a country where people will work for free?" (Delio, 2003).

In summary, for the first part of the chapter, we have built on two areas of our previous research and have set them in the context of the informational uncertainty generated by rapid information and communications technology (ICT) innovation and consumption. First, the debate on charging is dissected into political myths and funding strategies "to build capacity in an uncertain environment" (Longhorn and Blakemore, 2004, p. 16). Second, we are sensitive to the power relationships that ebb and flow in the PSI rights of access debates, for the positions and arguments in these debates mainly are "made by those groups who have most to gain (for example academically and commercially) from access to data" (Blakemore and Craglia, 2006, p. 21). To extend these considerations, two themes will now be addressed: Should data be free, and how are data made free?

The first theme postulates that the "free lunch is my right" — my taxes paid for the data, so I will not pay again, and as a citizen I have rights to see the data (subject, of course, to legislation such as privacy and data protection). On that basis, the case studies above of the U.S. Bureau of the Census and U.S. National Weather Service are indicative of a process of chipping away the range of data that are freely accessible to citizens, and this is what Harlan Onsrud terms the "destruction and despoliation of the public commons in information" (Onsrud, 1998). One of the most recent and influential articulations of this position was by the late Peter Weiss, whose "Borders in Cyberspace" (Weiss, 2002) is cited frequently as providing a rationale for free data, for example, in the 2006 U.K. campaign Free Our Data (Arthur and Cross, 2006a, 2006b), the Public Geodata* forum for Europe, or the U.K. Institute for Public Policy Research (IPPR) report on the relative balance between the protection of IPR and its commodified distribution. IPPR stated, at least, that politicians should critically examine their myths, perhaps in the U.K. moving toward a more public commons approach, noting that "policymakers should take account of the value generated by complementary products and services" (Pollock, 2006, p. 15). In fairness, it should be acknowledged that U.K. National Statistics did move radically away from chargeable data access to free data access in 2000 (Cook, 2000), and that the general guiding principles for access to PSI in the U.K. are evidenced in the Information

---

* http://publicgeodata.org/Home.

Fair Trader Scheme (IFTS) (HMSO, 2004) of the Office of Public Sector Information (OPSI), under which chief executives of government agencies are requested to

> make a personal commitment to the five principles
> for the re-use of Government information: openness,
> transparency, fairness, compliance and challenge.
> HMSO then examine the Trading Funds' underlying
> administrative and decision-making processes to ver-
> ify that they do in fact support the Chief Executive's
> commitment. (HMSO, 2003)

The Weiss study was a comparative analysis of what he saw as a predominantly European situation of protection of GI (IPR protection via copyright), and general policies of pricing GI to meet policies of cost recovery, or even semicommercialization. It should be acknowledged that the U.S. federal situation has never been to ban any cost, but to restrict charges only to what is termed the residual cost of dissemination, or "the sum of all costs specifically associated with preparing a product for dissemination and actually disseminating it to the public" (OMB, 1992) — charging only for the additional costs of making data available when that cost often now is near zero, with data being downloadable from the Web.

However elegant the arguments are, only a partial comparison is possible of the U.S. federal government to governments in Europe. It does not cover the bricolage of policies below federal level in the U.S., in state and local governments, which exhibit commodification and IPR protection. For example, the San Francisco Enterprise GIS* provides citizens** with access to a rich set of GI for the city. However, access to the data is via a registration page on which the terms and conditions for use must be accepted. These include "City and County of San Francisco does not charge for personal, non-commercial use of City spatial information," and any commercial use of the data must be with specific permission and under license arrangements. The license is very clear in setting out the terms of use. There is, at least, freely accessible use for nonprofit users, but the commercial focus shows that the gaps between European and U.S. reality are not as significant as Peter Weiss had stated. Similar terms of access and use operate in other U.S. states and municipal areas, too numerous to attempt to list here, as well as several variations, often within individual jurisdictions, whether at the state or local government level.

Indeed, the examples provided thus far show how the provision of something free almost inevitably occurs through resource provision using another

---

* http://www.sfgov.org/site/gis_index.asp?id=366.
** Mike applied online for access with his U.K. address and received the promised e-mail with access codes.

channel. We now explore a variety of free lunches to identify how the data or services have been made available for free. For example, in the Republic of Ireland, men and women over 60 years of age and disabled people are able to travel free on trains, a facility that is now extended across the border in Northern Ireland. From 2007, qualifying citizens from Northern Ireland will be able to travel free anywhere by train in the Republic, and vice versa (Hain, 2006). Two resource issues arise from what is a worthy social and political decision. First, if I am paying a full fare and am traveling on business, what is my reaction if the seats are fully occupied by those who are traveling free? Second, there are logistic irrationalities in any such scheme. Northern Ireland is part of the U.K. (and is therefore partner with England, Scotland, and Wales), but whereas citizens from the Republic of Ireland will be able to travel free to part of the U.K., citizens from the other three countries in the U.K. will not qualify for free travel in the Republic of Ireland. Free in this context seems therefore to mean differentially free, resulting in uncertain and exclusionary outcomes.

Free access to the Internet, particularly free broadband, was a frequently promoted claim in the U.K. from 2004 onwards. But, while access was free of charge, what were the particular terms and conditions? As Jane Wakefield warned early in the process, check for whether there are capacity limitations, e.g., charging after so much downloading or e-mail use, whether there is a fee to activate the service, whether technical support is available only via a premium-rate telephone service, and whether the free resource includes e-mail accounts (Wakefield, 2004). Similar concerns arose in Ireland when Internet access was first promoted (O'Hora, 1999), and in 2006, Google launched a free wi-fi service in Mountain View, CA, but the conclusions of a test were "it's not as reliable, as fast, or as easy to use, as my home internet connection or my cell phone" (Fehrenbacher, 2006). Free in this context therefore implies a restricted range of free resource, and to make up the package, other things are chargeable. Not surprisingly, therefore, user satisfaction with free broadband services in the U.K. fell in 2006 as "most providers fail to match rising customer numbers with improved services and technology" (BBC, 2006c).

The provision of a free resource may in itself generate uncertain outcomes that impose new costs. The provision of wi-fi hotspots in cafés has grown fast, and some cafés have started to provide free wi-fi access to attract customers. Some café owners found that some customers "would sit for eight hours purchasing a single drink, or nothing at all," and some customers even became angry when confronted with the fact that they were expected to buy drinks and food — after all, the wi-fi is free, so there cannot be an obligation to pay anything (Fleishman, 2005). Uncertain outcomes also influenced the development of free e-mail services such as Hotmail. As the use of free Hotmail expanded in the early 2000s, Hotmail developed payable services. Free accounts at one stage did not have automatic checking for spam e-mails; that was for the chargeable accounts only. A pricing motive to encourage a move from free to fee backfired when spam e-mail volume increased, and Hotmail ended up having to cope with huge volumes of e-mail (Olsen, 2002).

However, this experience did not stop Yahoo and America Online, in 2006, from proposing to charge e-mail accounts a fee not to have spam or junk emails delivered (AP, 2006). Even now, free Hotmail accounts only stay live if you log on within a set duration. In other contexts, the provision of free access to the Internet can produce market distortions. Initiatives to create wired cities such as New York (Wells, 2004) or Manchester (BBC, 2006g) are laudable in their attempts to maximize inclusion in the information society, but this comes with questions about whether access to the Internet is regarded as similar to public library provision (in which case you have to go to the library), whether the provision distorts market forces for other commercial providers, and who maintains and develops the infrastructure and resources (Grebb, 2005).

One interpretation of the above examples is that they are part of a genre of deconstructing a previously delivered full-service package, and then delivering what is regarded as the core or basic service that underpins societal needs. This model is particularly evident in the low-fare airline business. The previous definition of a flight with a full-service airline would be something like:

> Flight = Cost of {taxes, baggage allowance, baggage connection, airline flight from origin to destination, meals, compensation for delays, rerouting if connecting flights are delayed, etc.}

For a low-fare airline the definition is different, more along the lines of:

> Flight = Advertised cost of {airline flight from A to B (point-to-point connection only)} plus extra compulsory costs {government taxes and insurances} plus extra optional costs {meals, check-in baggage (the idea is that you take your baggage as cabin baggage, saving the airline the costs of employing baggage handlers, and therefore reducing turnaround time, and also making you the *de facto* baggage handler), food/accommodation problems if flights are delayed,* etc.}

Indeed, if you really carefully read the terms and conditions of airlines such as Ryanair, you will see that you also agree to donate up to 6 hours of your time to the airline on each flight. You are only liable for a refund of the fare if

---

* In 2006, the European Commission required carriers in the EU to provide more robust compensation for baggage loss and delays. One way, of course, to deter claims is to make the claims line accessible only via a premium rate telephone line.

the flight is cancelled or "is rescheduled so as to depart more than three hours before or after the original departure time" (Ryanair, 2006). That condition gives an airline flexibility to schedule flights at its most profitable convenience.

The prevailing theme with these examples is one of "if we cannot have it free, let's have it really cheap." A subliminal extension of "having it cheap" is one where the user of the service is taking the attitude "I want it cheap, and I assume that someone else will pay for the consequences of a low cost." This takes us into the area of ethics, and, more specifically, the ethics of consumption. Cheap airline flights in Europe are not paying for their contribution to greenhouse gas emissions and consequent climate change (BBC, 2006i). Here, the free lunch is leaving the cleaning costs for someone else to pay, and a U.K. government committee has considered introducing sales tax on air tickets, since travel has been zero rated for sales tax to date, as a form of environmental tax. Once a price is put on something like pollution, the problem itself can become a commodity that can be traded. Sulfur dioxide can be traded between companies, where one company that does not use its quota of pollution can sell the remaining quota to another company (Asaravala, 2004). Carbon taxes are sold and exchanged in the Chicago Climate Exchange, and the European carbon market could "trade $60 billion to $80 billion annually at a low price of $15 a ton" (Breslau, 2006).

On a more ethical level, the true cost of cheap clothing that we may buy means a well-being cost is passed on to those in poorer countries who work for low wages in squalid and dangerous conditions (Mathiason and Aglionby, 2006). Our selfish consumption in supermarkets, with cheap food that is available throughout the year, passes on a cost in terms of pollution, for example, through the "food miles" needed to fly fresh fish from Asia to Europe. Nearer to home our expectation that supermarkets will have stock that we want at short notice, for example, food and materials for a barbeque on a hot weekend day, means the supply chain needs to be highly controlled, often involving workers in distribution depots who work under extreme conditions of surveillance and control (Blakemore, 2005). The above examples therefore argue that the cheap lunch often involves costs that we pass on to someone else to pay indirectly. And, in some cases, the someone else can even be you, where the move to self-service checkouts in supermarkets means we do the work that before was the responsibility of a paid employee. We may be unpaid employees for newspapers when we send photographs (BBC, 2006a), and this activity itself generates ethical dilemmas — for example if a bomb goes off, do we take photos and send them to the media or help the injured (BBC, 2005a)?

The indirect passing on of cost is not always a negative experience, as the increasing provision of free online news and media content demonstrates. This activity will be explored in more detail later in this chapter, but the general approach to funding free content had been to rely on advertising revenue. A micropayment is given by an advertiser every time a hit occurs on a page with its advertisement (BBC, 2006b), and the pricing model for free online media content is indirect pricing, where the cost is covered by a donation of your time to view a paid advertisement. A variant development of this process is where you, as an individual, set a price by which advertisers can contact you with the advert. The Boxbe* e-mail service in 2006 aims to permit that approach, saying it "makes your inbox behave." Rather than having elaborate filters to remove spam and related emails, you decide who can send emails to your inbox and at what price, then "Boxbe will give 75 percent of funds collected from advertisers to users, who could optionally direct the money to a favorite charity" (Hudson, 2006).

However, a combination of click and pay is not always guaranteed to work, and has become subject to fraud as Google found out when it reviewed activity on its $6 billion a year advertising business. Fraudulent activity ranges from people clicking multiple times on a page, to writing programs to do the same, through to Trojan horse software that infects a PC and generates fake clicks (Schneier, 2006). A variant of the free media activity is evidenced with the U.K. British Broadcasting Corporation (BBC), where central state funding through a compulsory television license provides the BBC with significant funds to invest in digital media that are made freely available.** State funding may generate an unfair monopoly, and when the BBC was developing its digital media in 2001, there were fears from other commercial media outlets that they were being subject to unfair competition (Gibson, 2001; Trueman, 2002). There have been reactions to state monopoly of media channels in the past, notably in the U.K. during the 1960s with Radio Caroline*** and other "pirate" radio channels.

Free telephone calls are another form of free lunch. This seems wonderful, especially given our hyperconnected society, where we want to communicate, but where we are aware of the costs of international phone calls. The emergence of Skype was, to many, deliverance from the chargeable clutches of telecom companies. The Voice over Internet Protocol (VoIP) allowed people to communicate at no cost and is a classic example of disruptive innovation. But, as John Naughton warns, the service is not so much a free service as a service that uses peer-to-peer technology and utilizes your ICT resources. As he notes, there is a clear license agreement that you agree to when signing up for the service, where "Skype software may utilize the processor and bandwidth of the computer (or other applicable device) you are utilizing," admittedly only for the purpose of providing communication facilities for Skype users (Naughton, 2006). There are, however, examples of dramatically driving down prices using disruptive technologies such as VoIP. Hotxt, a

---

*   https://www.boxbe.com/ama/home.
** Indeed, we make unashamed use of the reliable, robust, and detailed free content of the BBC by citing it frequently where it provides useful examples and case studies. One by-product of the BBC's resources and dominance is that we can rely more realistically on the URLs being stable and the material being freely available.
***http://www.radiocaroline.co.uk/.

service launched in 2006 in the U.K., aimed at young people who send text messages. Instead of the usual 10 pence per message via conventional carriers, the cost was to be 1 pence (90% saving) since only the Internet carrying cost was charged (Economist, 2006b).

Free music was an emergent goal of users of the Napster peer-to-peer file-sharing service, which threatened to disrupt the copyright-controlled business of music sales. Again, it became a disruptive technology, rather like the introduction of reel-to-reel cassette decks in the 1970s, which allowed people to copy easily from a prerecorded cassette to a blank one. Or the introduction of photocopiers that allowed people to copy from copyrighted books and journals onto blank pieces of paper. Napster went from hero to villain during 2001–2002, when the music industry sought legal injunctions to force Napster to police the illegal copying of music (Zeidler, 2001), to its closure, through opportunistic business strategy when the pornography industry saw a benefit in the file-sharing technology (Zeidler, 2002), through to a relaunch of Napster in October 2003 when it became a legitimate music-selling business (BBC, 2004). Napster paved the way for later innovations in music distribution such as iTunes, which in February 2006 sold its 1 billionth music track (BBC, 2006e), "social machines" for photograph sharing and swapping (Roush, 2005b), and information-sharing applications such as Frappr for maps (Frappr, 2006).

3.6 Development, exploitation, and public investment

The information commons and the practice of information and knowledge sharing are at the heart of open-source software initiatives. Even here pricing is active, although the price of creating the software is written off by those working on the software, using a cost–benefit assumption that the benefits they receive in return are greater in value than the cost of their time. This argument is central to the knowledge-as-a-global-public-good view of Joseph Stiglitz, for quite apart from the expected economic benefits, e.g., more activity creates larger markets, which expands global economic activity, there is an ethical and moral consideration, for "it helps us think through the special responsibilities of the international community" (Stiglitz, 1998).

Provision of open-source software to developing nations, and strategic decisions to use such software nationally, involves a process of price and indirect costs. In late 2006, it was reported that "three quarters of UK colleges and universities adopt open source software" (Kablenet, 2006a), although there still is a price involved in free software, because the staff time involved in developing and supporting it is often regarded as a sunk cost and seldom is entered into the purchasing decision. There also are downstream potential cost implications, since the *Economist* reported that of the "roughly 130,000 open-source projects on SourceForge.net," no more that a few hundred still showed activity, and "fewer still will ever lead to a useful product" (Economist, 2006d). The counterargument would be that open-source activity is allowing considerably larger sharing of knowledge, leading to faster innovation cycles.

In some contexts, the need to pay for information can be seen as a form of discriminatory exclusion. For example, much academic writing occurs in journals that are expensive. A reaction to that is collaborative open-access journals, made freely available online (Dotinga, 2005). We made use of two open-access journals in our writing, *First Monday** and the *Journal of Digital Information.*** We wanted our writing to be available to the maximum readership, but the free access for readers was enabled through financial support, that is, the free lunch was subsidized for both magazines from large institutions. It seems unlikely that all academic writing could move to such a model, attractive though it is. A more extreme form of information exclusion is seen in developing nations, where severe limitations on resources mean they cannot afford to access the latest scientific literature.

Furthermore, as Florent Doiouf argued in 1994, this led to information imperialism, where people from beyond a developing nation publish research regarding that nation, "often developed in ignorance of the realities of life there, to make decisions with major consequences for all who live there" (Doiouf, 1994). Moves to address information imperialism include the Soros Open Society Institute decision in 2002 to invest in providing access to academic literature, and the decision of the U.S. National Academies Press (NAP) to make available scientific literature to over 100 developing nations in PDF format (Anon., 2004). In pricing terms, the NAP decision, though laudable, involves a very minimal residual cost of dissemination since dissemination is electronic. Making the information available for free does not lead directly to beneficial outcomes, as the United Nations Economic Commission for Africa (ECA) noted in 2005, when it requested that African governments move away from restrictive information and telecom practices and "commit themselves to policies that create information and knowledge economies" (UN, 2005). As Govindan Parayil noted, information can be available to overcome exclusions, but that intention can be confounded by "the unfair political economic context within which they are developed, deployed, and diffused" (Parayil, 2005, p. 49). In India, this requires government encouragement to not only use open-source software, but also change organizational and strategic behavior, since government departments very seldom invest in their IT resources, do not share their work, and the "government just sees free software as a way to save on licenses" (Thompson, 2006a).

The information and IT commons debate will, fortunately, continue to excite thinking, for such a debate is one of the only ways by which consensus can be achieved about the overriding principles of information and society. The developments noted earlier about Wikpedia conform to the view that the Wikipedia commons in 2006 may have moved from a free commons to

---

* http://www.firstmonday.org.
** http://jodi.tamu.edu/.

one that "offers only limited freedom maintained and controlled by an elite" (Klang, 2005). For the producers and owners of GI, however, the turbulent processes present ever more complex challenges. For example, as discussed in Chapter 2, what is the value of an information asset? This has resulted in an intangibles economy, noted earlier for information trading and futures, where the value of a company may not be invested in the previously traditional bricks and mortar, but the potential value of information and knowledge in future business.

For example, companies are raising capital by borrowing against the estimated market value of their copyrights, trademarks, and patents (Economist, 2006c). For the Ordnance Survey of Great Britain (OSGB), the valuation method used for their assets has resulted in a yearly, very public, disagreement with the government auditor. The annual turnover of OSGB is about £118 million, and the main income stream arises through what OSGB terms the "exploitation of data held in Ordnance Survey's National Geographic Database" (NGDB) (Survey, 2006a, p. 57). The creation of the NGDB has been funded over the years largely through public funding, and there is a question arising as to what is the value of the NGDB, since many knowledge businesses quantify their IPR as noted above. The OSGB has been consistently refusing to put a value on the NGDB in the annual financial returns, and the government auditor general has taken independent advice and classed it as an intangible fixed asset for which "the value to the business is not less than £50 million," and this then represents just under half of the overall fixed assets of OSGB (Survey 2006a, p. 57).

Why is this important? First, it reminds us that the value to the market is not just in the cost–benefits of using GI, but also in the potential investment in GI as a market in its own right. If OSGB were to be privatized, the initial public share offering (IPO) would need to be based on figures such as fixed assets and market potential. Second, it reminds us also that most government GI producers are not independent operators within their markets, but are operators whose activities are deeply constrained by government policies, and government policies are subject to sudden and unexpected change, just as the economy is subject to changes through the processes of globalization.

This finally returns us to the initial position that providing access to information is an economic and political contest between resource allocation and user demand. Just as we wrote, above, that the Treasury in the U.K. may be considering privatizing OSGB and other U.K. trading funds, we found out that the Office for Public Sector Information announced a partial shift in dissemination policy under which the Statute Law Database would now be available free of charge (BBC, 2007). The situation underpins the tensions between the politics of information and the economics of information. In the early part of the twenty-first century, these tensions have been exacerbated by global and local events such as 9/11 and global terrorism, globalization and mobility, and the emerging ability of the private sector to attack previously inviolable government data monopolies. With information, surveillance and monitoring (discussed in Chapter 5) are increasingly advocated by governments, who need to attack crime, defend borders, and deliver integrated services to citizens. Such integration activities stimulate more concern in citizens about privacy and data protection, thus adding more tensions to the economic and political contest.

References

Anon. 2004, April 5. The National Academies Provide Free Scientific Information to Developing Nations. National Academies USA. http://www4.nationalacademies.org/news.nsf/isbn/04052004?OpenDocument (accessed April 21, 2004).

AP. 2006, February 6. New Anti-Spam Tactic: Charge 'Em. Associated Press. http://www.wired.com/news/technology/1,70164-0.html (accessed February 6, 2006).

Archives. 2006, August 4. Discover Domesday. National Archives. http://www.nationalarchives.gov.uk/domesday/ (accessed August 4, 2006).

Arthur, C. and M. Cross. 2006a, March 9. Give Us Back Our Crown Jewels. *Guardian* (London). http://technology.guardian.co.uk/weekly/story/0,,1726229,00.html (accessed March 10, 2006).

Arthur, C. and M. Cross. 2006b, March 16. What Price Information? *Guardian* (London). http://technology.guardian.co.uk/weekly/story/0,,1731386,00.html (accessed March 16, 2006).

Asaravala, A. 2004, April 7. Using Capitalism to Clean the Sky. Wired.com. http://www.wired.com/news/business/0,1367,62968,00.html (accessed April 8, 2004).

BBC. 2002, May 7. Worldwide Weather Watchers Wanted. BBC. http://news.bbc.co.uk/2/hi/science/nature/1958296.stm (accessed November 16, 2006).

BBC. 2004, February 24. Napster Sells Five Million Songs. BBC. http://news.bbc.co.uk/1/hi/entertainment/music/3516167.stm (accessed February 24, 2004).

BBC. 2005a, August 5. Ethics Issue for Citizen Snappers. BBC. http://news.bbc.co.uk/1/hi/technology/4746633.stm (accessed August 5, 2005).

BBC. 2005b, March 24. New Online Access to 1871 Census. BBC. http://news.bbc.co.uk/1/hi/scotland/4376807.stm (accessed March 25, 2005).

BBC. 2005c, August 30. US Poverty Rate Continues to Rise. BBC. http://news.bbc.co.uk/2/hi/business/4198668.stm (accessed June 3, 2006).

BBC. 2006a, May 3. Amateur Snappers Get Their Reward. BBC. http://news.bbc.co.uk/1/hi/technology/4968940.stm (accessed May 3, 2006).

BBC. 2006b, April 6. Fewer Charges for Website Content. BBC. http://news.bbc.co.uk/1/hi/business/4880150.stm (accessed April 7, 2006).

BBC. 2006c, November 14. Free Broadband Users 'Less Happy.' BBC. http://news.bbc.co.uk/1/hi/uk/6145738.stm (accessed November 14, 2006).

BBC. 2006d, June 6. French Book Publisher Sues Google. BBC. http://news.bbc.co.uk/1/hi/entertainment/5052912.stm (accessed June 7, 2006).

BBC. 2006e, February 25. Itunes Achieves One Billion Mark. BBC. http://news.bbc.co.uk/1/hi/entertainment/4750744.stm (accessed February 27, 2006).

BBC. 2006f, July 11. Man Turns Paper Clip into House. BBC. http://news.bbc.co.uk/1/hi/technology/5167388.stm (accessed July 12, 2006).

BBC. 2006g, December 1. Manchester Plans Free City Wi-Fi. BBC. http://news.bbc.co.uk/1/hi/england/manchester/6199382.stm (accessed December 1, 2006).

BBC. 2006h, December 15. Obesity 'Could Bankrupt the Nhs.' BBC. http://news.bbc.co.uk/2/hi/health/6180991.stm (accessed December 15, 2006).

BBC. 2006i, August 7. Raise Air Travel Tax, Report Says. BBC. http://news.bbc.
co.uk/1/hi/uk_politics/5251022.stm (accessed August 18, 2006).

BBC. 2007, January 12. Government Looks at Data Shake-Up. BBC. http://news.bbc.
co.uk/2/hi/technology/6255321.stm (accessed January 12, 2007).

Blakemore, M. 2005, September. Surveillance in the Workplace: An Overview of
Issues of Privacy, Monitoring, and Ethics. GMB. http://www.gmb.org.uk/
Templates/Internal.asp?NodeID=92346 (accessed September 20, 2005).

Blakemore, M. and M. Craglia. 2006. Access to public sector information in Europe:
policy, rights and obligations. *Information Society*, 22: 13–24.

Blakemore, M. and R. Longhorn. 2001. Communicating Information about the World
Trade Center Disaster: Ripples, Reverberations, and Repercussions. *First Monday*,
6. http://firstmonday.org/issues/issue6_12/blakemore/index.html (accessed
December 10, 2001).

Blakemore, M. and S. Sutherland. 2005. Emergent commercial and organisational
charging strategies for geostatistical data: experiences disseminating UK official labour market information. *URISA Journal*, 16: 35–47.

Breslau, K. 2006, May 17. Pricing Pollution. *Newsweek*. http://www.msnbc.msn.com/
id/12839745/ (accessed August 3, 2006).

Brown, K. 2002. Mapping the future. *Science*, 298: 1874–1875.

Brown, M. 2006, December 2. Visitor Numbers Soar at Britain's Free Museums and
Galleries. *Guardian* (London). http://politics.guardian.co.uk/publicservices/
story/0,,1962329,00.html (accessed December 3, 2006).

Carroll, J. 2005, September 8. Katrina's Truths. *Boston Globe*. http://www.common-
dreams.org/views05/0905-26.htm (accessed September 13, 2005).

Ciborra, C. 2002. *The Labyrinths of Information*. Oxford: Oxford University Press.

Congress. 1974. Freedom of Information Act (Pl 89-487). U.S. Congress. http://www.
cni.org/docs/infopols/US.Freedom.Info.Act.html (accessed June 1, 2006).

Congress. 2005, April 14. National Weather Services Duties Act of 2005 (introduced
in Senate). U.S. Congress. http://thomas.loc.gov/cgi-bin/query/z?c109:S.786:
(accessed April 29, 2005).

Cook, L. 2000. Len On ... Free Data for All. *Horizons* (U.K. National Statistics), 15.

Cracknell, D. 2006, July 30. Government U-Turn on Free Information. *Sunday Times* (London). http://www.timesonline.co.uk/article/0,,2087-2291779,00.html (accessed July
30, 2006).

Crampton, T. 2006, January 21. A Battle of New vs. Not-So-New Tech. *International
Herald Tribune*. http://www.iht.com/articles/2006/01/20/business/voice.php
(accessed January 25, 2006).

Cross, M. 2006, December 14. Data Restrictions Cost Economy £500m. *Guardian*
(London). http://technology.guardian.co.uk/weekly/story/0,,1971028,00.html
(accessed December 14, 2006).

DCA. 2006. *Draft Freedom of Information and Data Protection (Appropriate Limit and Fees)
Regulations 2007*. 47. Department for Constitutional Affairs, London.

Delio, M. 2003, June 6. Jobs Squeeze for Indian Workers. Wired.com. http://www.
wired.com/news/business/0,1367,59126,00.html (accessed June 9, 2003).

Doiouf, F. 1994, October 20. A sermon on information imperialism. In Annual Conference of the International Association of Aquatic and Marine Science Libraries and Information Centers, Honolulu. http://www.irf.org/irinfgis.html
(accessed January 22, 2003).

Dotinga, R. 2005, April 11. Open-Access Journals Flourish. Wired.com. http://www.
wired.com/news/medtech/0,1286,67174,00.html (accessed April 12, 2005).

Dunion, K. 2006, June 21. Decision 104/2006: Mr. Brian Smith and the Assessor for Tayside Valuation Joint Board.Scottish Information Commission. http://www.itspublicknowledge.info/appealsdecisions/decisions/Documents/decision6104.htm (accessed August 24, 2006).

Economist. 2005a, September 15. How the Internet Killed the Phone Business. http://www.economist.com/opinion/displaystory.cfm?story_id=4401594 (accessed September 16, 2005).

Economist. 2005b, September 15. The Meaning of Free Speech. http://www.economist.com/business/displayStory.cfm?story_id=4400704 (accessed September 16, 2005).

Economist. 2006a, February 2. Dividing the Pie. http://www.economist.com/finance/displayStory.cfm?story_id=5468383 (accessed February 3, 2006).

Economist. 2006b, March 30. Hot to Trot. http://www.economist.com/business/displayStory.cfm?story_id=6750257 (accessed April 1, 2006).

Economist. 2006c, June 15. Intangible Opportunities. http://www.economist.com/finance/displaystory.cfm?story_id=E1_SDGRPRN (accessed June 18, 2006).

Economist. 2006d, March 16. Open, but Not as Usual. http://www.economist.com/business/displayStory.cfm?story_id=5624944 (accessed March 17, 2006).

Economist. 2006e, August 24. Who Killed the Newspaper? http://www.economist.com/opinion/displaystory.cfm?story_id=7830218 (accessed August 25, 2006).

Fehrenbacher, K. 2006, August 7. Search of Google Wi-Fi. BBC. http://news.bbc.co.uk/1/hi/technology/5251646.stm (accessed August 10, 2006).

FGDC. 2004a. Guidelines for Providing Appropriate Access to Geospatial Data in Response to Security Concerns. 13. FGDC, Washington, DC.

FGDC. 2004b. NSDI Future Directions Initiative: Towards a National Geospatial Strategy and Implementation Plan. 10. FGDC, Washington, DC.

FGDC. 2006. NSDI Cooperative Agreements Program. FGDC. http://www.fgdc.gov/grants (accessed February 14, 2006).

Fleishman, G. 2005, June 14. Cafés Find Wi-Fi Boom Unsettling. International Herald Tribune. http://www.iht.com/articles/2005/06/13/business/wifi.php (accessed June 20, 2006).

Floweb. 2006, March. Flight Tracking Services. Floweb.com. http://www.floweb.com/antest/ge/intro.aspx (accessed March 26, 2006).

Frappr. 2006, March. About Frappr. Frappr.com. http://www.frappr.com/ (accessed March 13, 2006).

GAO. 2001. 2000 Census: Significant Increase in Cost Per Housing Unit Compared to 1990 Census. General Accounting Office, Washington, DC.

GAO. 2004. 2010 Census: Cost and Design Issues Need to Be Addressed Soon, GAO-04-37. General Accounting Office, Washington, DC.

GAO. 2005. Army Corps of Engineers Lake Pontchartrain and Vicinity Hurricane Protection Project: Statement of Anu Mittal, Director Natural Resources and Environment: Testimony before the Subcommittee on Energy and Water Development, Committee on Appropriations, House of Representatives, GAO-05-1050T. Government Accountability Office, Washington, DC. http://www.gao.gov/new.items/d051050t.pdf (accessed April 21, 2007).

GeoConnections. 2006. Geoconnections Annual Report 2005–2006: Laying the Groundwork. 29. GeoConnections Division, Ottawa. http://www.geoconnections.org/publications/reports/ar/05-06_AR_E.pdf (accessed April 20, 2007).

GfKRoper. 2006. 2006 Geographic Literacy Study. 89. National Geographic Society, Washington, DC.

Gibson, O. 2001, November 7. BBC Presses Ahead with Controversial Online News Service. *Guardian* (London). http://media.guardian.co.uk/newmedia/story/0,7496,589224,00.html (accessed November 7, 2001).

Google. 2006, February. Google Earth Enterprise: Overview. Google. http://earth.google.com/earth_enterprise.html (accessed February 14, 2006).

Grebb, M. 2005, October 19. Cities Unleash Free Wi-Fi. Wired.com. http://www.wired.com/news/technology/wireless_special/0,2914,68999,00.html (accessed October 20, 2005).

Hain, P. 2006, June 26. Secretary of State Announces All-Ireland Free Travel Scheme for Senior Citizens. Northern Ireland Office. http://www.nio.gov.uk/media-detail.htm?newsID=13256 (accessed July 10, 2006).

Haines, L. 2005, September 13. Google Earth Threatens Democracy. *Register.* http://www.theregister.co.uk/2005/09/13/google_earth_threatens_democracy/ (accessed September 13, 2005).

Haines, L. 2006, July 29. Chinese Black Helicopters Circle Google Earth. *Register.* http://www.theregister.co.uk/2006/07/19/huangyangtan_mystery/ (accessed August 24, 2006).

Harris. 2002, June 25. Harris Corporation Awarded $200 Million Contract for U.S. Census Bureau's MAF/Tiger Accuracy Improvement Project. Harris Corporation. http://www.harris.com/view_pressrelease.asp?act=lookup&pr_id=952 (accessed June 1, 2006).

Harris. 2006, January 17. Harris Corporation Awarded $40 Million Contract Extension by U.S. Census Bureau for Maf/Tiger Program. Harris Corporation. http://www.harris.com/view_pressrelease.asp?act=lookup&pr_id=1750 (accessed June 1, 2006).

HMSO. 2003, September 4. *Review of the First Four IFTS Verifications.* HMSO, London.

HMSO. 2004. *Information Fair Trader Scheme.* HMSO, Cabinet Office, London. http://www.opsi.gov.uk/ifts/ (accessed April 9, 2007).

Hudson, J. 2006, December 13. Startup Makes Spammers Pay. Wired.com. http://www.wired.com/news/technology/1,72288-0.html (accessed December 13, 2006).

Kablenet. 2006a, August 8. Buy None Get One Free. Kable Government Computing. http://www.kablenet.com/kd.nsf/Frontpage/C3AAE73CB791A0E8802571C40 03F6BF1?OpenDocument (accessed August 8, 2006).

Kablenet. 2006b, December 14. DCA Presses for FOI Changes. Kable Government Computing. http://www.kablenet.com/kd.nsf/Frontpage/EC287E3CE84E52F 580257244003D73B0?OpenDocument (accessed December 14, 2006).

Klang, M. 2005. Free Software and Open Source: The Freedom Debate and Its Consequences. *First Monday*, 10. http://firstmonday.org/issues/issue10_3/klang/index.html (accessed March 9, 2005).

Klapperich, C. 2006, July 25. Excessive Fees Violate Information Act. *Northwest Herald.* http://www.nwherald.com/print/286641065771907.php (accessed August 24, 2006).

Lash, S. 2002. *Critique of Information.* Sage, London.

Longhorn, R. and M. Blakemore. 2004. Re-visiting the valuing and pricing of digital geographic information. *Journal of Digital Information*, 4: 1–27. http://jodi.tamu.edu/Articles/v04/i02/Longhorn/ (accessed March 7, 2007).

Martin, L. 2006, June 18. Wikipedia Fights Off Cyber Vandals. *Guardian* (London). http://observer.guardian.co.uk/uk_news/story/0,,1800273,00.html (accessed June 18, 2006).

Mathiason, N. and J. Aglionby. 2006, April 23. The True Cost of Cheap Clothing. *Observer* (London). http://observer.guardian.co.uk/business/story/0,,1759240,00.html (accessed April 23, 2006).

McGee, K. and J. Skågeby. 2004. Gifting Technologies. *First Monday*, 9. http://first-monday.org/issues/issue9_12/mcgee/index.html (accessed January 29, 2005).

Mosco, V. 2004. *The Digital Sublime: Myth, Power and Cyberspace.* MIT Press, Cambridge, MA.

Naughton, J. 2006, June 11. Telecoms Pray for Time When the Skype Finally Falls In. *Observer* (London). http://observer.guardian.co.uk/business/story/0,,1794639, 00.html (accessed June 11, 2006).

NRC. 2003. *Weaving a National Map: Review of the U.S. Geological Survey Concept of the National Map.* National Research Council, National Academies Press, Washington, DC.

ODPM. 2004, July 21. Ordnance Survey (Framework Document). ODPM. http://www. publications.parliament.uk/pa/cm200304/cmhansrd/cm040720/wmstext/ 40720m01.htm#column_13 (accessed July 21, 2004).

OFT. 2006, December 7. OFT Report Finds Public Sector Bodies Cost the Economy Half a Billion in Hidden Information Markets. OFT. http://www.oft.gov.uk/ News/Press+releases/2006/171-06.htm (accessed December 14, 2006).

O'Hora, A. 1999, November 1. Free Internet Access: What It Costs. Bizplus.ie. http:// www.bizplus.ie/bp_online/e-business/?ns=24 (accessed June 17, 2002).

Olsen, S. 2002, March 21. Microsoft Sweeps out Hotmail Accounts. CNET News. com. http://news.com.com/2100-1023-866086.html?tag=cd_mh (accessed March 22, 2002).

OMB. 1992. *OMB Circular a-130, Management of Federal Information Sources.* OMB, Washington, DC.

OMB. 2005. US Geological Survey: National Cooperative Geological Mapping. OMB. http://www.whitehouse.gov/omb/expectmore/summary.10003722.2005.html (accessed August 25, 2006).

Onsrud, H.J. 1998. The tragedy of the information commons. In *Policy Issues in Modern Cartography*, D.R.F. Taylor (Ed.). Pergamon, Oxford, pp. 141–158.

Parayil, G. 2005. The digital divide and increasing returns: contradictions of informational capitalism. *Information Society*, 21: 41–51.

Pollock, R. 2006. *The Value of the Public Domain.* 18. Institute for Public Policy Research, London.

PricewaterhouseCoopers. 2000, March 9. PricewaterhouseCoopers Projects $11 Billion Loss in Federal Funds to State and Local Governments if Census 2000 Is Undercounted. http://www.pricewaterhousecoopers.com (accessed March 30, 2000).

Richtel, M. 2006, December 13. Skype to Charge for Calls to Landline and Mobile Phones. *International Herald Tribune.* http://www.iht.com/articles/2006/12/13/ business/skype.php (accessed December 13, 2006).

Roush, W. 2005a, May. The Infinite Library. *Technology Review.* http://www.technology-review.com/articles/05/05/issue/feature_library.asp (accessed April 14, 2005).

Roush, W. 2005b, August. Social Machines. *Technology Review.* http://www.technolo-gyreview.com/articles/05/08/issue/feature_social.asp (accessed July 18, 2005).

Ryanair. 2006, August. Terms and Conditions of Travel. Ryanair.com. http://www. ryanair.com/site/EN/conditions.php (accessed August 31, 2006).

Schaefer, M. and J. Moeller. 2000, March. Support for International Infrastructure Activities. Federal Geographic Data Committee. http://www.fgdc.gov/international/sha1.pdf (accessed May 5, 2005).

Schneier, B. 2006, July 13. Google's Click-Fraud Crackdown.*Wired Magazine*. http://www.wired.com/news/columns/1,71370-0.html (accessed July 13, 2006).

Shachtman, N. 2004, June 18. Galileo: Challenge to U.S. Might? Wired.com. http://www.wired.com/news/technology/0,1282,63865,00.html (accessed June 18, 2004).

Shaker, L. 2006. In Google We Trust: Information Integrity in the Digital Age. *First Monday*, 11. http://firstmonday.org/issues/issue11_4/shaker/index.html (accessed April 20, 2006).

Sternstein, A. 2005, October 17. Mapping Technology Threatens USGS Jobs: Fewer Mapmakers Needed as Agency Prepares a Competitive Sourcing Bid. *Federal Computer Week*. http://www.fcw.com/article91117-10-17-05-Print (accessed October 20, 2005).

Sternstein, A. 2006a, August 29. Ransom Payments Set Maps Free. *Federal Computer Week*. http://www.fcw.com/article95833-08-29-06-Web (accessed September 7, 2006).

Sternstein, A. 2006b, July 3. Senators Give Mapping Center a Chance for Survival. *Federal Computer Week*. http://www.fcw.com/article95133-07-03-06-Web (accessed September 7, 2006).

Stiglitz, J.E. 1998. Knowledge as a Global Public Good. World Bank. http://www.worldbank.org/knowledge/chiefecon/articles/undpk2/w2wtoc.htm (accessed July 19, 2000).

Survey. 2001, December 20. Mapping Agency 'Should Have More Freedoms and Responsibilities': Government-Owned Company Status Proposed for Ordnance Survey. Ordnance Survey. http://www.ordsvy.gov.uk (accessed December 20, 2001).

Survey. 2006a. Annual Report and Accounts 2005–06. 76. Ordnance Survey, Southampton, U.K.

Survey. 2006b. Trading on Geographic Intelligence: How Ordnance Survey Benefits the Nation. Ordnance Survey. http://www.ordnancesurvey.co.uk/oswebsite/media/features/tradingfund.html (accessed February 16, 2006).

Thompson, B. 2006a, May 12. India Lays Down 'Open' Challenge. BBC. http://news.bbc.co.uk/1/hi/technology/4764565.stm (accessed May 12, 2006).

Thompson, B. 2006b, August 25. Not as Wiki as It Used to Be. BBC. http://news.bbc.co.uk/2/hi/technology/5286458.stm (accessed August 25, 2006).

Treasury. 2006. *Pre-Budget Report. Investing in Britain's Potential: Building Our Long-Term Future*. HM Treasury, London.

Trueman, P. 2002, May 30. Working the Web: BBC. *Guardian* (London). http://www.guardian.co.uk/online/story/0,3605,724213,00.html (accessed May 30, 2002).

UN. 2005, April 26. UN Commission Calls on Africa to Commit to Information and Knowledge Economy. UN News Service. http://allafrica.com/stories/200504270003.html (accessed April 29, 2005).

USGS. 2001. *The National Map: Topographic Mapping for the 21st Century*, final report. U.S. Geological Survey, Washington, DC.

Wakefield, J. 2004, March 4. Hidden Costs of Budget Broadband. BBC. http://news.bbc.co.uk/1/hi/technology/3529989.stm (accessed March 4, 2004).

Weiss, P. 2002, February. Borders in Cyberspace: Conflicting Public Sector Information Policies and Their Economic Impacts. Department of Commerce. http://www.primet.org/documents/Weiss%20-%20Borders%20in%20Cyberspace.htm (accessed March 16, 2005).

Wells, M. 2004, August 21. New York Set for Citywide Wireless. BBC. http://news.bbc.co.uk/1/hi/technology/3578982.stm (accessed August 24, 2004).

Wired. 2004, December 15. Prepping to Pull the Plug on GPS. Wired.com. http://www.wired.com/news/technology/0,1282,66056,00.html (accessed December 16, 2004).

Withers, D. 2005, May 29. Santorum's Bill Would Restrict Public Access to Weather Service Data. *Chicago Tribune.* http://www.kansascity.com/mld/kansascity/news/politics/11768740.htm (accessed June 5, 2005).

Zeidler, S. 2001, March 2. Napster's Demise Causes Panic. Reuters. http://www.washtech.com/news/media/8055-1.html (accessed March 3, 2001).

Zeidler, S. 2002, September 13. Pornographer Offers to Buy Napster. Reuters. http://www.reuters.co.uk/newsArticle.jhtml?type=internetNews&storyID=1444192 (accessed September 13, 2002).

chapter four

Pricing information:
The interaction of mechanism and policy

4.1   *Pricing theories*

The examples provided in Chapter 3 illustrate that in recent years the pricing of information and information technology goods has been subject to considerable volatility. This section explores the extent to which prevailing theories of pricing can help explain the examples covered in previous sections. The first area of theory, known as price discrimination, addresses how prices can be set given a demand from a segmented market, and is categorized in three sections based on the original definition by Pigou in 1932 (Wikipedia, 2006).

4.1.1   *First-degree price discrimination*

First-degree price discrimination is where the producer sells the same goods to different market segments at different prices. The determinant is the ability or willingness of the customer to pay a price (Dedeke, 2002). For example, IGN Belgium* sells its topographic data at the scale of 1:10,000 for different prices, depending on the type of area**: rural areas cost half the price of urban areas; i.e., in the 2006 price list, from 10 to 40 euro per square kilometer in rural areas (depending upon area size purchased), compared to 20 to 80 euro for urban areas. The selling of cars has classically been a first-degree pricing process. There is an advertised or recommended price from which discounts are given for large fleet purchasers or selectively for individual customers through trade-in discounts and special offers. The opaque nature of new car pricing has historically made it difficult for potential purchasers to effectively discriminate between vendors. Such customer uncertainty has now encouraged some manufacturers to move from variable to fixed pricing. However, this also can generate problems, as when U.S. car manufacturer Ford announced a "clear pricing strategy"; what was meant to say "'Here's a justifiable and reasonable price' can come across in ads as 'Hey, we won't rip you off this time!'" (Mahoney, 2006).

---

* http://www.ngi.be/.
** http://www.ngi.be/FR/FR1-5-1-1.shtm.

4.1.2   Second-degree price discrimination

Second-degree discrimination focuses more on volume discounts, but where the volume prices are the same for all (Varian, 1996). The range of prices in the IGN Belgium example above also demonstrates second-degree price discrimination. The greater the geographical area for which you purchase data, the lower is the price per square kilometer, e.g., the cost for urban data declines from 80 euro per square kilometer for coverage up to 20 square kilometers to 20 euro per square kilometer for coverage in excess of 100 square kilometers.

Dedeke splits this category further into three subcategories. First, is the conventional volume discount approach. The second is features-based, for example, where there is "deactivation of several functions of a software product that is being sold to a special category of customers" (Dedeke, 2002). This is frequently used in commercial software packages and information services, including geographic information systems (GISs) and GI-based online information services. A reduced subset of a product or service is made available free of charge, which the vendors hope will then encourage people to pay for the full-service product. U.K. householders can use no-cost, partly deactivated services to check for potential flood risk, possible pollution risks, and the value of nearby properties using services such as Landmark (2003b), Sitescope (2003), Nethouseprices* (2005), and even the U.K. Environment Agency** (Environment, 2003).

Dedeke's third category of price discrimination is the time-based approach, for example, where a video shop charges more for a new release DVD than for an old film. The best example of this in the GI world is access to meteorological data from those government agencies who do charge for such information, for example, the U.K. Met Office. Under special arrangements, much raw meteorological observations data is available for free (for noncommercial use, cost of distribution only), mainly for education and research, once a certain period has passed, which may vary from 24 hours to days or weeks. The point is that the most valuable weather data are used for immediate and short-term forecasting, for which there exists a proven and very active marketplace, e.g., a 1-year license to use the U.K. Met Office national 24-hour forecast on a single website cost £515 in 2007.***

A geographical variant of this type of price discrimination is where flat rates are charged for a service irrespective of the distance covered, but there is a volume discount. The one-price charge by Amazon.co.uk for delivery

---

* See Benedictus (2005) for a discussion of the possible privacy implications of the ready availability of property prices and details.

** This example brings into focus the issue, discussed elsewhere in the chapter, of whether the launch of a commercial service by a government agency is unfair competition against services provided in the commercial sector. Note also the similarity between this example and the weather data debate, noted earlier, in the U.S.

***http://www.metoffice.gov.uk/newmedia/datafeed/catalogue.html.

means that the same price is paid by customers whether they live next to the Amazon warehouse that dispatches their order or 600 kilometers away in the north of Scotland. The flat rate is varied depending on the amount purchased and whether a faster delivery mode is selected, but the resulting charge is still the same irrespective of distance. The Amazon pricing model for product dispatch combines a form of fixed-price universal service for sending an order to a customer who has selected the products online, where Amazon benefits from the "connectivity and low transaction latency" of the Internet (Odlyzko, 2004, p. 341).

4.1.3   Third-degree price discrimination

Third-degree discrimination focuses more on the ability of market segments to pay, discriminating, for example, between low-ability groups such as elderly people and students, and high-ability groups such as the urban affluent. That means the U.K. Met Office, Britain's government meteorological agency, continues to provide weather data for the public good (the traditional weather forecast is in the public commons*), but the Met Office then has a series of commercially available value-added services that are targeted at specific sectors. For example, forecasts of icing on airplanes allow airports to plan de-icing more cost-effectively (Met Office, 2005). The Met Office site** lists a range of other services, such as long-range local forecasts for people taking out insurance against the cancellation of outside events due to bad weather, and services for supermarkets so that they can plan to have the optimum food stocks in place in stores — there is little logic in stocking lots of barbeque food for a weekend that will be washed out by wind and rain.

The Ordnance Survey of Great Britain (OSGB) applies this type of pricing to what it terms licensed partners.*** In this category, the price of using OSGB data is constructed from a fixed plus variable cost. The fixed cost covers the administrative costs of maintaining the relationship and providing the data and support. The variable cost is a revenue stream that is a proportion of the sales price of the value-added applications undertaken and marketed by partners. In an extensive review of information in the global economy and society, Scott Lash differentiates between information that is sold as a commodity (exchange value) and that which generates added value through reuse and repackaging (use value), warning that in the information society much more revenue is generated through use value (Lash, 2002). In the context of GI we could apply those criteria to OSGB, where Lash would warn

---

* That means you hear the weather forecast free on the radio or television and can look for local forecasts on sites such as http://www.bbc.co.uk/weather or globally on sites such as http://www.weather.com.

** http://www.metoffice.gov.uk/services/index.html.

***http://www.ordnancesurvey.co.uk/oswebsite/partnerships/licensedpartners/index.html.

that just selling data is not likely to generate significant income where the market is increasingly complex and requires more sophisticated data production (Longhorn and Blakemore, 2004). Put simply, the ongoing use value of OSGB data is an increasingly important part of the income stream, and this then imposes increasing demands on OSGB for support and database development, including updating, maintenance, and enhancement of the data itself.

The pricing model of OSGB is a hybrid approach to a complex market, building in particular on the importance of use value. The revenue stream is made up of core customer groups who pay a central license fee that covers all users in the sector, e.g., utilities, local government, and academia are core license areas. In addition, a form of universal service license existed for some years in the context of the National Interest Mapping Services Agreement (NIMSA). This covered costs for central government usage of the data and guaranteed that areas where a revenue-focused business would not concentrate resources, such as remote rural areas, coastlines, etc., would continue to be mapped to the same resolution, timeliness, and quality as other areas in Britain (DCLA, 2006). However, like the museums example earlier in this book, NIMSA was contingent on the willingness of the U.K. government to pay a central subsidy. Late in 2006, after a review of the costs and benefits, the agreement was terminated. OSGB's reaction was to advise users that there would be "an impact on the currency and content of the rural geography in our products" (Survey, 2006b), with the possibility of longer rural revision cycles. However, the contest between supply and demand was evident again in the statement by OSGB that it would explore technological efficiencies, i.e., doing more for less cost, in order to try to compensate for loss of the NIMSA funding, and that key activities such as data for emergency services and coastal mapping would be maintained "in the national interest despite the extra cost burden."

Strategic national interest developments can also be funded on a public–private basis, such as the initiative to produce large-scale underground asset three-dimensional maps (Kablenet, 2006). There are then revenue streams from sales of printed products, from licensing data to private sector companies, from value-added partnerships with private sector companies, and from OSGB's own commercial digital products (Survey, 2006a).

4.2   Extending pricing theory

Using the range of examples of the free lunch discussed in Chapter 3, we propose to extend the levels of price discrimination to include zero-degree price discrimination. This category is primarily concerned with the pricing of public sector information goods, where pricing mainly is set through a public subsidy that allows the organizations to disseminate the data largely free of any charge. The pricing dilemmas that emerge in this context are articulated by Claudio Ciborra, including how to avoid "free riders," such as ourselves, for example, when we applied for access to the San Francisco data in the example in Chapter 3, and "who should pay for the positive and/or negative externalities created by use?" (Ciborra, 2002, p. 60).

4.2.1 Zero-degree price discrimination

In the zero-degree price discrimination category are organizations such as most U.S. federal agencies or U.K. National Statistics. In these organizations, the data owner has no ability to link the market take-up of data to any reinvestment program, other than to beg for more subsidy funding. This makes for a nice scenario on the basis that data are free to all users, and superficially, the data are easily disseminated via the Internet, i.e., friction-free with no replication costs beyond the initial sunk investment. Yet there is no mechanism by which user needs can be linked to the funds that will satisfy them, for example, for new data formats or new types of data. In recent years, the new public management approaches have allowed the naïve belief that efficiency gains will deliver service improvements. However, whatever happens, the zero-degree price discrimination category initially mediates data to users, as in "Here it is, it's free, use it!" but then dis-intermediates potential service improvements from customer needs, i.e., "Well, it's free, so don't come to us asking for more!"

Central planning approaches to government have long since been criticized politically (for example, communism), but the more plausible reason for the zero-degree category being so problematical at present is that the mechanisms by which governments obtain income have moved substantially from direct taxation to indirect taxation and user charges. With smaller proportions of the total population entering the labor market, which impacts directly on levels of direct (income) taxation, as well as political imperatives to lower the levels of taxation (to keep the voters happy), and with more people living into old age, resulting in greater demands on health and social services, the political attraction of indirect taxation is significant for politicians desperate to satisfy all sections of the voting public. Even the elderly pay sales tax, and the need for government to temper financial demands on health services can be offset in part by the customers paying for some services.

More frequently, a form of rationing of the service is used, known as the waiting list — you can have the treatment, but you will need to wait some time. The same can be seen with zero-degree GI. The current less-than-complete state of U.S. national topographic mapping (NRC, 2003) was substantially the result of historical underinvestment, exacerbated by the fact that there was no income stream other than the government central subsidy. The Weaving The National Map approach has been an attempt to appeal to national altruism, through cooperative agreements (FGDC, 2006) as a means of indirectly funding improvements in national mapping. It says in effect, "Let us work together to weave all the high-quality data held at various geographical levels," but the major cost of doing this is to be borne by the data owners at state and local government levels. Furthermore, it was no surprise that, given the inability of the U.S. Geological Survey to maintain the maps, new public management techniques would be used, e.g., competitive tendering was planned for many of the USGS activities (Sternstein, 2005), as well as altruism through the "active participation and support by the geospatial community at all levels" (USGS, 2005a, p. 5) via "sustainable partnerships" (USGS, 2005b).

The National Map* accessible through the Geospatial One-Stop strategy (USGS, 2004) would provide coordinated and centralized access to national mapping data and was considered central to the delivery of government programs. The Government Accountability Office (GAO) noted that Geospatial One-Stop is a high-risk and critical project (GAO, 2006). It is not surprising that the initial strategy was not to throw huge amounts of money at USGS to update its mapping, but to see if a collaborative national map could be built. The U.S. federal government has demonstrated a historical underinvestment in data, and now is demonstrating a realization that the cost of updating information is significant. For example, the National Flood Map Modernization Coalition wrote to the Office of Management and Budget in July 2005, regarding mapping undertaken by the Federal Emergency Management Agency (FEMA). They were concerned that the budget requests by FEMA over many years had not been met by the level of federal grants, and the important flood insurance rate maps had become out of date. The task of updating the maps from 1996 onwards required significant levels of investment (NFMMC, 2005, p. 2). In the context of zero-degree pricing, therefore, the initial free lunch rather defers, to a later stage, the costs of reinvestment.

Under the conditions of zero-degree pricing, an organization — inevitably a government organization — will spend much of its time trying to match growing demand against finite funding. In the U.S. situation, this is further exacerbated by the inability to generate extra funding due to the constraints imposed by federal policy to make information available for free, both of charge and of copyright (OMB, 1990, 1992, 1995, 2002). The doctrine of free information has been debated at length and is covered elsewhere in this book, but the basic arguments go like this. If we make data freely available, then it stimulates more economic growth. With more economic growth, more businesses will employ more people and will generate more taxes. The increase in taxation income will be greater than the cost of creating and maintaining the data. The doctrine in the past was semireligious in its fervor, and largely assumption-led, but started to unravel when the economy downturned and government revenues declined or were transferred to other priorities.

As is the case throughout history, warfare and, more recently, global terrorism provide temporary respite for funding fears, by increasing military spending and increasing investment in surveillance technologies (Dotinga, 2004; Ward, 2004; Webb, 2004; Willard, 2005), which can directly benefit the GI

---

* http://nationalmap.gov/.

and GIS industries. These violent events can also backfire on the government and the industry, as was the case in the U.K., where fears about the European Union INSPIRE directive liberating and integrating geographic information (Rennie, 2006) caused totally unfounded paranoia about terrorists being able to predict the movement of submarines. Also, war and national security are by no means guarantees that additional funding will be provided for national or international mapping work. This is evidenced by statements from the American Geological Institute's Government Affairs Program senior policy advisor, John Dragonetti, in May 2002, stating that "another issue of concern is the lack of funding for USGS's significant activities in support of homeland security and the overseas war on terrorism. All four divisions of the USGS have been heavily involved in national security but neither the emergency supplemental appropriations passed last fall nor the FY 2003 budget provide funds directly for these activities" (Dragonetti, 2002).

4.2.2   The consequences of underfunding national map production

Even where fear or paranoia or concerns over homeland security do generate additional funding, these gains are often only a temporary respite from the underlying endemic problem of demand outstripping supply. In the end, it still comes down to funding. Indeed, in the 2005 report on the U.S. National Map Project, potential partners were questioned and the dominant response was to say that funding assistance is needed (USGS, 2005a, p. 98). Therefore, the collaborative program is in effect a piecemeal process of indirectly purchasing the data for the national database.

The result of zero-price discrimination can be seen at its most extreme in Egypt, where the lack of strategic investment by government in national mapping at the Egyptian Survey Authority (ESA) is apparent from the poor state of what should be its primary resources. One of ESA's legal responsibilities is for the boundaries in the national cadastral system, especially in rural areas — a different ministry is responsible for the title details. ESA clearly does not update these maps frequently, and the land registration information has no effective update process in place. The existing update system for maps is unstructured, and there is even a "lack of an agreed practice manual" (Elrouby et al., 2005, p. 1), although this was being addressed in a new initiative that started late in 2005. The lack of updated mapping goes back to 1921, when a report into the state of mapping noted that as of February 1919 (Egypt, 1921), 44% of the maps were over 15 years out of date, 75% were more than a decade out of date, and 11% were 3 years out of date — and these were the most current that existed. The present rural cadastre maps, most of which comprise inked boundary changes on the original paper maps created in the late 1930s to mid-1940s, are presently being digitized, relying on a dual-level subsidy of government money and significant contributions through foreign aid projects. The quality of the final digitized rural cadastral database — a legally binding data set under Egyptian land registration law

— will be questionable from the outset due primarily to decades of underinvestment in primary mapping activities.

There is no significant national-level provision of updated topographic mapping in Egypt, and like in the U.S., the market has responded by creating its own products. Potential customers have often taken outdated ESA maps and used them as a base on which to build their own internal data holdings. Furthermore, not only has a large proportion of the potential market decided not to wait for ESA to produce quality data, but the market also is forming collaborative alliances and portals to publish, disseminate, and market their data in information portal services (Tamima, 2006). As a result, topographic mapping in Egypt is produced as a bricolage of data that are all held beyond the control of the national mapping agency, as the following examples indicate:

The Egyptian Gas Company (GASCO) reports that it uses what it terms "a high accuracy reference network" and that the 1:50,000 ESA maps are used as a backdrop to its own high-accuracy data (Geovision, 2002).

The Greater Cairo Utility Data Center carries out its own survey activities to produce 1:5,000 base maps, using GPS, for its own infrastructure data needs. Thus, one of the biggest potential customers for ESA data seems to be collecting its own information and is developing added-value services that will rival ESA's offerings (Cairo, 2004; Sayyed Badr, 1997).

Egypt Post. In the 2005 edition of the National Information Society strategy, there is note of a public–private partnership between Egypt Post and Federal Express. Such a development would need substantial base mapping, yet the current supply situation would seem to force such initiatives to go to the private sector for more updated information at less ground precision (MCIT, 2005, p. 70).

The Egyptian Antiquities Information System has been building its own GI holdings. We were also informed that it had been paying ESA to do surveys. More importantly, its website is very clear about the fact that it has updated ESA maps on its own, and so would not likely be a customer of ESA now (EAIS, 2006).

Central Agency for Public Mobilization and Statistics (CAPMAS) has its own GIS center, which has as one of its tasks "establishing 1:5,000 scale digital infrastructure maps for all governorates of Egypt with all required codes" (CAPMAS, 2006). Since 1993, CAPMAS produced its own maps at 1:5,000, covering Cairo, Alexandria, and the Canal Zone.

Connection. There is now a private sector company that has commercialized CAPMAS products, called Connection. Connection provides digital mapping information at scales of 1:50,000, 1:25,000, and 1:5,000; other products include building footprints (Connection, 2006).

EgyMaps. Furthermore, Connection now partners with the private sector GI specialist Quality Standards Information Technology (QSIT), in an Internet portal called EgyMaps, which will deliver route-find-ing tourist online maps, advertising links, and business location services (EgyMaps, 2006). EgyMaps and Connection are the sorts of data provision and service development functions that would use-fully serve the Egyptian Geography Network (EGN) (QSIT, 2005) in the absence of complete and updated ESA information.

Vodafone. The telecoms sector is developing its own detailed topo-graphic data for Egypt, including roads, demographics, and related data that allow service planning. Vodafone is also looking at returns on investment through value-added services using the data (Voda-fone, 2003, p. 8).

The key conclusion from these examples is that zero-discrimination pric-ing, in effect central subsidy from general tax revenues, is a remarkably difficult pricing regime within which to build market-relevant data. Further-more, it can lead, by default, to a form of creeping privatization where the actual geographic information infrastructure data for a nation is collected, processed, disseminated, and used beyond any realistic influence from gov-ernment. This certainly is the case in Egypt, and a similar case exists in the U.S. for large-scale GI — larger scale than the 1:24,000 USGS topographic coverage of America, which itself is not fully up to date.

4.3 *Pricing contexts: issues*

Other information pricing approaches focus more on pricing contexts. The pricing issue here involves managing the relationship between the price charged for the information and the time it takes to obtain the information. For example, Snyder differentiates between a pricing strategy that must recover all the costs of the organization (absorption) and one that needs only to recover part of the costs (contribution). The latter is familiar to public sec-tor GI as the residual cost of dissemination approach (Snyder and Davenport, 1997), where an organization is only able to charge for the additional costs of making the information available, one of the charging-related best-prac-tice principles included in the pan-European INSPIRE SDI directive. This is supported by Hughes, arguing: "The average cost per unit of information will continue to decline, but that the share of revenue taken by application rather than content will rise" (Hughes, 2001, p. 10). Therefore, there is logic in OSGB capitalizing on use value by developing its own value-added prod-ucts as well as licensing partners to do the same. This, however, then gen-erates fears among those partners of perceived, or real, market distortions through unfair exploitation by OSGB of its own intellectual property rights (IPR) (OPSI, 2006) when there is no viable or realistic competitive data supply available to licensed partners that would encourage price competition.

There are, however, limits to the ability of a producer to ask for a share of the onward profits from use value. For example, the provider of avocados to a Michelin Star restaurant is highly unlikely to expect a percentage of the restaurant profits where the avocado is used in a meal. First, there is competition between producers of avocados, and while all avocado producers could conceivably form a cartel to demand a percentage, the restaurant could respond by removing avocados from the menu. Second, a Michelin Star restaurant would be sourcing the best and highest-quality avocados, and these will be sold at higher prices. By comparison, geographic information frequently has been produced only in one form, leading often to near *de facto* monopolies, whether state-owned mapping agencies or private firms.

Others look more at the channels through which information can be disseminated, combined with types of information. For example, advertising revenue has been one model through which the new media industry in particular attempted to fund free availability of their content. They took advantage of the fact that use of the Internet minimized the distribution costs almost to zero (Schiff, 2003), although the two major weaknesses here were the inability to match income stream to user demands, and an underestimate of the costs of maintaining the archive of content. Bates and Anderson look at product quality and completeness as a means of allowing differential pricing, in particular in helping to discriminate between what is free and what is not; for example, reliability, authority, update, aggregation and integration, full selection, and flexible download, all of which give the customer a high value-to-cost ratio (Bates and Andersen, 2002).

Shapiro adds to this criteria of product differentiation and personalization the use of promotions to lock in customers to your service and, where there is competition, clearly differentiating your product from others (Shapiro and Varian, 1999). However, product differentiation is challenging in the context of the new global reach of companies and the overall neutrality (OECD, 2006, p. 4), i.e., homogeneity of channel distribution via the Internet. Hughes looked at the likely consolidation of three major information players — Factiva, Dialog, and Nexis — noting that the Internet distribution channel presents a paradox. A small player can enter the market at relatively low cost, but needs to fight against the dominant profile enjoyed by large players, and that in turn requires a higher innovation rate, which in turn generates higher levels of turbulence and uncertainty in the market (Hughes, 2001).

The impact of uncertainty was noted by Evans and Wurster (2000) in a review of the turbulent experience of the *Encyclopaedia Britannica* when it was moving from print to online format in the face of competition from diverse Internet information sources, not least from other, perhaps less known, encyclopedias.

Lastly, Evans and Wurster advise of two issues that were becoming even more pronounced in 2006. First, the longer the reach of your business, the more likely it is to encounter "asymmetries of information — differences in knowledge among people or companies that affects their bargaining power" (Evans and Wurster, 2000, p. 38). Second, the turbulences of the global information market mean that there will be more deconstruction, "the dismantling and reformulation of traditional business structures" (Evans and Wurster, 2000, p. 39).

Since the mid-1990s, the information market has been affected significantly by factors such as globalization trends, swings in national and international markets, the growing potential customer base enabled by increased Internet access, and the dynamics of IPR protection. As a result, pricing strategies have become both complex and turbulent, often changing according to market situations and rapidly emerging competition.

4.4 Market positions and roles

Within this turbulent pricing strategy, what is the market position of the producer? What can the producer do to try to increase market share at least cost and least risk? This section looks at several issues that can impact on market positioning for GI products and services.

4.4.1 First mover advantage

An important factor influencing the information producer is first mover advantage, i.e., being the first to launch a new information genre or product and to have the market reach that allows rapid take-up, typically by access to adequate venture capital for sales and marketing or via enlightened marketing policies. Examples here include the Arc/Info GIS, via which the Environmental Systems Research Institute (ESRI) employed first mover advantage by encouraging use of its GIS in the higher-education sector, a strategy used many years ago also by Apple Computer in the U.S. in relation to its early microcomputer products. Skilled GIS students then move into both the public and private sectors, taking their skills and knowledge with them. Why should employers pay for new training when they have access to a large cadre of GIS staff trained at the public's expense in university?

ESRI demonstrated strategic awareness, for example, when, in the late 1980s, Michael Blakemore was technical advisor to the Economic and Social Research Council's Regional Research Laboratories Initiative (RRL) (Masser and Blakemore, 1991). Blakemore approached U.K. GIS vendors to explore whether discounted provision of GIS could be negotiated, arguing that the RRLs were strategic research centers that would produce high-quality GIS experts. No such luck with the U.K. vendors, but the U.S. company ESRI was more than happy to agree to beneficial license terms, and the RRLs became one of the conduits through which Arc/Info became the dominating GIS in the U.K. In the area of e-commerce, eBay experienced a combination of first mover advantage, for an innovative new business method, and significant availability of risk capital, allowing for new products and services to be launched globally. A similar situation exists today for Google Earth and its copycat services.

4.4.2   Avoiding legacy systems problems

Another factor increasing the likelihood of success involves avoiding or overcoming legacy factors. In this context, being able to launch new products without having to continue to service legacy customers who still need support for already existing information styles or formats is a distinct advantage. It is best seen in more social contexts, where dating sites and virtual communities have been quickly successful, but where there also is a high "churn rate." A typical legacy issue in GI was exemplified by the decision of a mapping agency to improve positional accuracy of its product using GPS, a logical enhancement of the product. However, the legacy-related costs were then borne mostly by the customers, who had to rework all their applications to cope with the increased accuracy and resolution. While most users were happy to receive GI of higher quality than before, many were also unhappy with having to bear this unexpected — and mainly unbudgeted — additional expense, some felt without adequate prior consultation. Paradoxically, the improvements in accuracy are fundamental to the opening of new applications, such as the micromanagement of vineyards using high-resolution maps and GPS (AP, 2004), so one person's economic benefit can be another's economic cost.

4.4.3   Enjoying, protecting, or abusing a monopoly position

A further factor for success in the marketplace is having some form of monopoly power. It is perfectly acceptable for a business to protect its activities by exercising its intellectual property rights (IPR), whether these relate to copyright, database protection (mainly in Europe), or patenting of business methods or algorithms (mainly not in Europe). A patent is a time-constrained monopoly on the exploitation of the patent holder's intellectual property in the device, method, process, or algorithm covered by the patent. IPR can be quite useful, as the legal dispute between Landmark plc and Sitescope plc showed in 2004, when Landmark won a legal case for "infringement of its copyright in Home Envirosearch, the market-leading environmental report for homebuyers" (Landmark, 2003a). The fear of monopoly exploitation also threatens the relationship between commercialized government trading funds such as OSGB and commercial resellers. In a judgment between OSGB and the company Intelligent Addressing in July 2006, the U.K. Office for Public Sector Information (OPSI) agreed that the actions of OSGB in licensing data to Intelligent Addressing breached rules of openness, transparency, and fairness in data availability terms and pricing (OPSI, 2006).

A process monopoly would be typified by the Egyptian Survey Authority (where property can only be registered using its maps), the U.K. rail system (where route monopolies are granted to franchisees), or the enforced relationship that occurs between GB local government, the utilities, and the Ordnance Survey through the legislative requirement to use OSGB data for planning purposes or for land registration. In many U.K. hospitals, the provision of television and telecoms services to patients is undertaken by a monopoly provider. The costs to patients are significantly in excess of costs to domestic providers — up to 15 times for a telephone call (Kablenet, 2005) — and there is concern that this represents monopolistic exploitation of captive customers.

Monopoly behavior can occur even where there is an apparent competitive market with multiple suppliers. In 2005–2006, the U.K. government carried out an inquiry into the cost of rail fares (Commons, 2006), not only observing that the privatized rail companies have *de facto* route monopolies, but also receiving evidence that there is a strong similarity between fares for competing transportation modes; for example, the first-class rail fare from Newcastle to London was nearly the same as the business airfare for the same route. A further monopoly practice has been seen in the practice of some hotels to charge unreasonably high prices for Internet access (Taylor, 2006), when other hotels offer free access. Those who charge high prices seem to follow a pricing practice of enforced lock-in, where the opportunity costs of going outside the hotel to find cheaper Internet access are too high. Those offering free access will be expecting higher consumption of drink and food through room service as the customers work for longer periods in their rooms. Lastly, in 2006, there was concern that there may be pricing collusion between the suppliers of online music (Economist, 2006c), and this was particularly worrying at a time when the music publishers were increasingly effective at reducing illegal downloading and sharing of copyrighted music.

Sometimes a government data monopoly can be weakened by the poor nature of the data themselves, and by inflexible pricing and dissemination policy, and business can contribute information back to government in ways that government cannot. The growth of geodemographics, with global companies such as CACI, Equifax, and Experian, occurred because business did things that government could not. They classified data and, by so doing, made subjective statements about the socioeconomic characteristics of locations.

Experian Business Strategies (EBS),* part of the global group that builds geodemographic and credit-referencing profiles, has been marketing a value-added service using U.K. government official employment statistics, formerly the Census of Employment and now the Annual Business Inquiry (ABI).** EBS takes the official statistics and does things that government official statisticians cannot do, either because they are organizationally not capable of doing something, or because it is not permitted under their professional standards of work. First, EBS will model one cycle of data against another to check for anomalies. Historically, the government statisticians were not obliged to do this and processed each survey as if it were new. Second, EBS can model the

---

* http://www.business-strategies.co.uk/Home.aspx.
** http://www.business-strategies.co.uk/Products%20and%20services/Economic %20forecasting/Making%20sense%20of%20the%20ABI.aspx.

surveys against other data, can interpolate where there is missing data, and can provide estimations for geographies other than the official geographies. These are all processes that are not generally undertaken within the official statistician code of statistics. Both central and local government agencies subscribe to the Experian service, which therefore provides government with an arms-length mechanism to add value to official data in ways that they cannot or would not normally do.

A further Experian service has been developed, forecasting trends for European regions.* This is based on the Eurostat** Regio statistics. Eurostat is bound by the legislation of official European Union statistics. It must wait for member state agencies to provide data. It can only process data according to official rules of harmonization and to official EU geographies. It can only process data for the EU member states, which means that it provides data for French colonies, but not for Norway. Experian, by contrast, can acquire individual country data as soon as they are released for use, can combine EU data with non-EU data to provide pan-European coverage, can interpolate and forecast, and can value add in ways not permissible for the official statistical agency of the EU.

Lastly, monopolistic behavior has been emerging rapidly through the worrying patenting of ideas or business methods, a process that runs strongly counter to the conventions of not taking out copyrights on ideas.*** Examine the patents taken out by Multimap in 2001 (USPTO, 2001a, 2001b), which relate to "displaying the locations of one or more places — hotels, restaurants, stores, etc. — on a map, with hyperlinks between the map and pages of information about the location" (Multimap, 2001). Look at the U.S. National Security Agency patent in 2005: "Patent 6,947,978, granted Tuesday, describes a way to discover someone's physical location by comparing it to a 'map' of Internet addresses with known locations" (McCullagh, 2005). Then become very worried about a large range of U.S. patents in the area of geographic information handling (USPTO, 2005a, 2005b, 2005c, 2005d, 2005e, 2005f, 2005g, 2005h). The *Economist* has described this process as in intellectual arms race, where the outcome may be "mutually assured destruction," the MAD scenario of the old superpowers arms race, where "companies amass patents as much to defend themselves against attacks by their competitors as to protect their inventions" (Economist, 2005a). The patenting of ideas leads to two forms of disruption to the market, both involving what has become known as patent trolls (Kintisch, 2006). First, there are trolls that are companies who exert their patent by threatening smaller companies, who then

---

* http://www.business-strategies.co.uk/Products%20and%20services/Economic%20forecasting/European%20Regional%20Service.aspx.

** http://epp.eurostat.ec.europa.eu/.

***In 2006, Dan Brown, best-selling author of the novel *The Da Vinci Code*, successfully defended himself against claims of other authors that he had stolen the idea. The U.K. High Court ruled that ideas cannot be patented, but as we see here, this process is alive and well in the geographic information sector.

either pay up because they lack the resources to fund a legal fight, or close down, thus removing innovative activity from the market. Second, there are trolls who have managed to patent an idea and then challenge large companies who have used the idea on the basis that it has been, and is still, in the public domain. For example, trolls have attacked Microsoft, diverting resources "that should have gone to advancing technology to make functionality go quicker, better, cheaper" (Kintisch, 2006).

4.5   *Pricing contexts: costing mechanisms*

What is the pricing context? Here the context sets the ground rules for the strategy. Examples include subsidy costing, contribution costing, absorption costing, and indirect costing:

Subsidy costing: The cost of the service is underpinned by a flat-rate payment from government. This is the classic pricing position of the free-data believers. A loss leader version of subsidy occurs when a product is launched at a low price, and then increases once users are locked in, a favorite approach for magazines that will run for a set number of weeks. Another variant is predatory pricing, where prices are depressed below cost to undercut a competitor.

Contribution costing: Focuses on the behavior of costs rather than their function. This aspect involves cost recovery, or the contributory aspects of data sharing. For example, in the proposed production of U.K. identity cards, both private and public sectors see benefits in sharing resources so that the police would be alerted if someone they were seeking used an identity card in the purchase of a commercial service (Hinsliff, 2006).

Absorption costing: All costs are to be covered by pricing. This is typified by cost recovery, which is the most basic form of absorption, i.e., cover your costs. In the U.K., this model is extended by the trading funds, where a data producer must recover all its costs plus a percentage extra that is returned to the National Treasury.

Indirect costing: The running costs are paid by an indirect income stream that has no direct relationship to the costs of service provision. Advertising revenue that covers the costs of free-access media sites was the most common example in the early 2000s. It does, however, suffer from a critical weakness in that there is no direct control over the matching of income to expenditure. Around 2001–2002, when there was a global economic downturn, the mismatch of resources led to considerable instability in the newspaper industry, for example, with the *New York Times* reducing activity and staffing levels for its Web content (Krebs, 2001).

Many other contexts help in the setting of prices. Similar products can be differentiated, such as "own brands" in supermarkets that may be manufactured by producers whose personal brand products are sold in the same store at a higher price. Versioning is used frequently in software, through regular upgrades, and this also provides a mechanism to lock in customers, since the upgrade and maintenance prices of software are much lower than the original purchase price, thus dissuading people from switching to competing software.

4.5.1   Time dependency in pricing

Price can be time dependent. If you are a serious investor in the stock market, you will want the latest share prices, since global markets can move within milliseconds,* and the services that provide you with the information are priced at a premium. If you are a casual investor, then you can rely on the large range of free price services, but the trade in value is time. Most prices will be 20 minutes old, which is of no particular consequence to a casual investor, but is seriously outdated for stockbrokers. Time and demand interact in pricing that is demand based, with the high-profile users of this approach being low-fare airlines. It is also used by U.K. train services or hotels in many cities, where prices increase when demand is highest, and that includes socially important times such as weekends and Christmas, when cheap rail or air fares are often hard to find (Webster, 2005), or when cities are hosting major conferences or sporting events, and cheap hotel rooms disappear. Such pricing approaches are even being experimented with for musical performances. Hitherto the price of a particular seat has been set in advance, and the price is charged whether the customer books 1 year or 1 week ahead. Auction approaches can be initiated with the Internet booking systems, asking customers how much they would be willing to pay for a seat (Walker, 2003). A more nefarious version of demand pricing is where cartels emerge to increase prices across the board at high-demand times. Christmas 2005 in the U.K. saw accusations that "leading electronics companies have been accused of ramping up prices for online shoppers in the run-up to Christmas" (BBC, 2005a).

4.5.2   Impact of payment strategies and technologies

Use the power of the Internet to focus on the marginal costs of processing a payment. There should be adequate income left after the administrative costs of processing the payment have been deducted from the payment received. These are known as micropayments (Thompson, 2006). Before online financial transaction services became available, the cost of processing a small payment often was excessive, so retailers would only accept check payments. In the context of micropayments, this pricing/charging/payment regime led to the growth of intermediary financial services such as Paypal (Paypal, 2003),

---

* See http://www.rba.co.uk/sources/stocks.htm for the range of stock price sources.

with eBay and Google more recently moving into the areas of "electronic wallets" (AP, 2006). Such a development encourages a move away from direct relationships between price and cost of service, toward a strategy that builds, encourages, and enables mass usage of products or services. It now becomes ever more feasible to sell 1000 items (especially digital items) at 1 euro and still earn a profit, due to reduced overhead costs for payment collection, versus having to sell 10 items at 100 euro to be profitable, with much of the additional per-product revenue disappearing in administrative costs.

4.5.3   Strategies that circumvent pricing

Some strategies may try to circumvent price; for example, making information available in a form that predates a chargeable form. This is used increasingly by academics who are aware that the high-cost journals in which they publish are read by a limited range of people. Copyright law does not allow the authors to make the final published paper available, except by prior agreement with the publisher, who will in most cases hold sole or joint copyright in the article. However, a preprint version of the article often may be distributed more freely. In this instance, the published information is degraded in order to make it available in the commons more quickly, e.g., the preprint version may not be as complete or as authoritative as the final published document. The outcome of this process, as noted by the *Economist*, is that making information available before it has gone through a peer group evaluation, "helped to keep the scientific process accurate," because errors and misunderstandings can quickly be disseminated (Economist, 2003) online versus the typically long delays between submission and publication in peer-reviewed journals.

At a higher level, academic funding bodies, such as the Research Councils U.K., proposed to mandate free access to all research that they have funded. In principle, this is a logical move to make available the outputs to the maximum audience, but they had to reduce demands when they "met with stiff opposition from traditional journal publishers" (Wray, 2006). The same experience occurred in the U.S., where the National Institutes of Health (NIH) spend nearly $30 billion a year on research and were willing to spend between $2 million and $4 million a year to create and support an electronic archive that would make research outputs freely available. Their proposals were weakened after commercial publishers complained that their business would be threatened, and also, complaints were received from "professional societies that fund their activities by publishing journals" (Economist, 2005b). Such tensions encouraged the European Commission to initiate a "study on the economic and technical evolution of the scientific publication markets in Europe" (Europe, 2006) in July 2006. In the above contexts, the proposals to create a free lunch did not fully take into account the interconnected business models that were represented by the pricing of the information products. A hybrid model emerged in September 2006, as Google announced that it was making available, free of charge, 200 years of global newspapers, where "free and charged-for articles are displayed side-by-side" — at least offering the customer an opportunity to try free-of-charge content before deciding to pay money (BBC, 2006a).

The dynamic interplay of increasing information availability, the uncertain use and integration of that information, and strategic reactions from information owners converge in the emergence of online comparison shopping sites, a typical one being Amazon.com Marketplace and Google's Froogle site (Schmidt, 2003). From the convenience of a computer, customers can now shop around in a way that avoids the time and expense incurred when walking or driving from physical store to physical store. Some services provide more than lists of prices, and Pronto was launched as an application that monitors over 50,000 online stores, monitoring the searching activity of the customers "until it finds a better deal. Then it sends a message prompting the user to click away" (Tedeschi, 2006). Comparison shopping goes beyond geographical borders, as demonstrated in July 2005, when the BBC reported that for selection of IT products the online price in the U.K. was 3.5 times the online price in U.S. Web stores (BBC, 2005b). Such is the volume of small packages being ordered directly from the U.S. that U.K. customs could not intercept more than a tiny proportion and charge import duty.

4.6   Changing relationships between information producers and users

The dynamic interplays are therefore generating "radical re-conceptualizations of the roles of and relationships among content creators, intermediaries, and consumers" (Slater et al., 2005, p. 4). Instead of using the traditional intermediaries such as agents and publishers, new authors can use low-entry-cost services such as Lulu to publish their writing online, set the access and price terms, and then release the book online (BBC, 2006b; Lulu, 2005). Traditional publishers are reacting against the physical and online intermediary sellers (the "bookshops") by selling directly in competition with online retailers like Amazon.com (Goldfarb, 2005). What is happening is that the boundary between information and services is increasingly blurring, so it is no surprise that information producers are aggressively moving into service provision, or that service providers are building or buying into information resources themselves.

4.6.1   Producers and service providers fight back

The fluid interplay of service and information is seen with hotel companies and airlines, where the emergence of intermediary online agencies such as Expedia and Opodo took customers away from hotel and airline sites. A combination of better sites, better customer loyalty strategies, and assurances that the best prices were to be found on their own sites has meant the hotels and airlines "have gained control of online sales despite fears years ago that independent Web sites would take the majority of business" (Peterson, 2006b) — a process of remediation. Online intermediary agencies are therefore finding new innovations to retain customers in an environment where their excellent integrated information offerings are used by customers to compare broad costs, but who then book directly online with the supplier (Peterson, 2006c). A further response to that development is a more intelligent disintermediated travel portal. The service Travel Meta Search raised over $10 million in 2006 to establish a service that will "search airfares, hotels, car rentals, and vacation packages from both mainstream and discount airline sites" (Schenker, 2006). In this context, the interplay of information, innovation, and business strategy means that customers need ever more sophisticated informational skills to take the best advantage of rapidly changing market offerings.

4.6.2   *Paying for exclusivity and protecting the brand*

Price may be related not to the traditional component costs of production, but to a higher value based on brand and exclusivity. Hermes, Versace, and other designer clothing and fashion goods attract prices that are well beyond their component costs, and this price is protected in part by attempts to prevent forgeries and fakes of their products. However, take a walk from St. Mark's Square in Venice, past the Café Florian on your left, and proceed down the narrow street of exclusive stores. The premium brand shops are there, but outside them on trestle tables are groups of North African traders selling fake versions of the products you can see in the shop windows. Real and fake live side by side, and you can observe the Italian police walking past the "IPR criminals." So something more than legal enforcement is going on, and the high price in effect states that only the very rich can afford the products. Owning the real thing makes you part of a select community, and you are therefore less worried by the person opposite with a fake bag, for it reinforces your exclusivity. This premium pricing is at its most complex with expensive watches, where the Patek brand, with watches retailing at $1 million, vets potential purchasers to ensure that they are not just speculators: "It's almost like the customer has to apply to be an owner" (Gomelsky, 2006).

Exclusivity combines price with premium service. One of the problems encountered when we focus exclusively on the provision of free resources is that we often do not see the widening gap between those with free access and those who pay. Just as we celebrate the availability of low-fare airlines, and the ability of more people to travel, we often fail to see the exclusive, high-cost services moving ever more distant from us. Lufthansa in 2004 decided to segregate first-class travelers from the rest of traveling humanity in a new terminal dedicated to premium travelers (Lufthansa, 2004). Such elite people receive personal attention and are taken to the airplane by limousine. Lufthansa is one of the airlines that has targeted premium, or extreme, travelers, and such people "care less about free award tickets or upgrades than about getting the sort of personal service" (Peterson, 2006a) they desire, i.e., they simply travel too much to want to travel more on free tickets. The distancing of the elite from mass consumption spaces is not new and exists also in the consumption of literature. In the U.K., the working class, typified as ignorant and uneducated, was in many cases literate and knowledgeable. Jonathan Rose found that when the intellectual elite saw that the literature they were specialized in reading was also read and discussed by the masses, they had to find a new literature genre to colonize intellectually, "like a genteel household that moves to ever more remote suburbs, to escape the crowds of the encroaching inner city" (Rose, 2001, p. 438). This process is nicely parodied in the satirical magazine *Private Eye* in their "Psueds Corner"* section, where opaque and obscure text is ridiculed.

The brand and exclusivity pricing approach is a form of reputation pricing, wherein the preservation of reputation is ever more challenging with the global flows of information. A worker at Buckingham Palace, the home of the U.K. queen, was sacked for trying to auction one of the Queen's Christmas puddings on eBay in 2004 (Reuters, 2004). The *Encyclopaedia Britannica* examples earlier in the chapter show how fee and free compete, and Britannica needs to preserve its price levels by justifying and maintaining its reputation. In 2006, the journal *Nature* published a comparison of "errors" in Britannica and Wikipedia, and Britannica saw it necessary to provide a detailed and very public rebuttal of the allegations (Economist, 2006a). There are very real dangers to established brands caused by rapid global dissemination of bad news, partly because there are now such low-cost opportunities for launching an attack against the brand — a practice at which many conservation-oriented nongovernmental organizations (NGOs) excel when attacking the latest oil company, logging firm, or genetic engineering research lab. Complaints blog websites are easily and rapidly established, where customers can share their bad experiences about low-fare airlines (Bowes, 2006)** or allegedly poor cable Internet and television services.*** Protecting brand, price, and market position is an increasingly complex task, which with growing environmental and ethical awareness "will also have to signal something wholesome about the company behind the brand" (Economist, 2001).

---

* http://www.private-eye.co.uk/sections.php?section_link=pseuds_corner&.

** And intriguingly in this customer-centric society there are commensurately fewer instances where customers establish websites that share praise and good experiences — bad news, as ever, travels faster than good news, and (just read any newspaper) bad news has a higher market value.

***http://www.nthellworld.co.uk/home.php.

4.7 Conclusion

As an overall conclusion to both Chapters 3 and 4, there is considerable turmoil in the content industry, whether it is public or private sector. The newspaper industry continues to exemplify attempts both to lock in customers and to find ways of funding the cost of service provision through indirect and direct charging (Robinson, 2006). Give newspapers away by relying on advertising revenues (Economist, 2006b). Make information available online only after a certain time has passed, or try it the other way, where current news is free online but the archive is chargeable (Graybow, 2005; Seelye, 2005). "Buy" information about the customers by making them register for online access (Dvorak, 2004), but understand that you have little ability to authenticate the identity of the user: "Depending on my mood, I'm a 92-year-old spinster from Topeka whose hobbies include snowboarding, macramé and cryptology" (Penenberg, 2004). Try something and quickly evaluate it, such as the *Economist* in November 2006, which introduced a requirement for nonsubscribers to acquire a day pass to access content: "click below to view an advertisement and then proceed to Economist.com's premium content," yet within weeks that process had disappeared from the website. Bite the bullet and tell people that online charging is now in operation (Murphy, 2004; Independent, 2003). Or search for something radically new, that moves away from the characteristic where "Internet journalism is still largely material from old media rather than something original" (Crosbie, 2004).

The growing tendency for GI data producers to charge for their data is nothing special or new, when set into the context of the Internet-era content industry. While there are still laudable examples of attempts to build the Commons, for example, Open Street Map (Openstreetmap, 2006) and similar citizen-based mapping projects, it presently seems unlikely that such initiatives will seriously threaten established players; that is, until they reach a critical mass — as did Wikipedia in 2006 — and then have to look seriously at formalized structures that need proper resourcing. This wide range of examples indicates that there is growing, not reducing, economic, political, and social turbulence in the pricing of information, and that free lunches will continue to be experimented with, but will be subsidized centrally only with difficulty, no matter what the emotional and economic arguments are about justification and need.

The choices range from the free lunch comprising a cheese sandwich and pickle, e.g., incomplete, 25-year-old topographic data at medium scale from an underfunded mapping agency, to the nonfree lunch comprising a three-course dinner with wine, e.g., the mapping agency that has reinvented itself as a geospatial resource center along commercial, information market industry lines, offering fully digital data resources updated 50,000 times a day. Then there are the partially subsidized lunches at either extreme. The point we wish to make is that different diners have different appetites (data and service requirements), different lunchtime budgets (which might also vary over time and circumstance), and different sponsors (subsidizers). Evidence indicates that the GI market is continually evolving to take all these variations into account, while continual innovations in the geospatial technology industries, as well as information processing and delivery industries, ensure that ever more options will arise in the future for producers and users.

References

AP. 2004, April 13. Winemakers Get Juiced about Tech. AP. http://www.wired.com/news/technology/0,1282,63047,00.html (accessed April 14, 2004).

AP. 2006, June 29. Google vs. Ebay: Who's Gonna Pay? AP. http://www.wired.com/news/technology/1,71275-0.html (accessed June 30, 2006).

Bates, M.E. and D. Andersen. 2002, April. Free, Fee-Based and Value-Added Information Services. *Factiva.* http://www.factiva.com/index.asp?node=menuElem0 (accessed May 17, 2002).

BBC. 2005a, November 15. Firms 'Ramping up Online Prices.' BBC. http://news.bbc.co.uk/1/hi/business/4438298.stm (accessed November 16, 2005).

BBC. 2005b, July 20. Web Shows High Costs of Hi-Tech. BBC. http://news.bbc.co.uk/1/hi/technology/4700445.stm (accessed July 21, 2005).

BBC. 2006a, September 6. Google Opens up 200 Years of News. BBC. http://news.bbc.co.uk/2/hi/business/5317942.stm (accessed September 7, 2006).

BBC. 2006b, March 8. Online Publisher's European Move. BBC. http://news.bbc.co.uk/1/hi/business/4782018.stm (accessed March 10, 2006).

Benedictus, L. 2005, February 18. Street Wise. *Guardian* (London). http://www.guardian.co.uk/g2/story/0,,1417108,00.html (accessed February 21, 2005).

Bowes, G. 2006, June 11. Websites Vent Travellers' Spleen. *Observer* (London). http://observer.guardian.co.uk/travel/story/0,,1794692,00.html (accessed June 11, 2006).

Cairo. 2004, June 24. Future strategy for execution work management through work programs: practical study. In *Transforming Government through Technology Summit*, Cairo Governate, Sister City Program of the City of New York. http://www.nyc.gov/html/unccp/scp/downloads/pdf/cairowhitepaper.pdf (accessed February 8, 2006).

CAPMAS. 2006, February. Central Agency for Public Mobilization and Statistics. CAPMAS, Egypt. http://www.capmas.gov.eg/ (accessed February 8, 2006).

Ciborra, C. 2002. *The Labyrinths of Information.* Oxford University Press, Oxford.

Commons. 2006, May 10. Sixth Report: How Fair Are the Fares? Train Fares and Ticketing.House of Commons, Transport Committee. http://www.publications.parliament.uk/pa/cm200506/cmselect/cmtran/700/70002.htm (accessed June 5, 2006).

Connection. 2006. Connection. Al-Alamia Company for Programming and Information Systems. http://www.connectionmaps.com/about_us.htm (accessed February 8, 2006).

Crosbie, V. 2004, March 17. Weak Online Economics Threaten Quality of All Journalism, Pew Study Finds. Online Journalism Review. http://www.ojr.org/ojr/business/1079553393.php (accessed May 5, 2004).

DCLA. 2006. *The National Interest Mapping Services Agreement (NIMSA),* Annual Report 2005–06. 17. Department for Communities and Local Government, London.

Dedeke, A. 2002. Self-Selection Strategies for Information Goods. *First Monday,* 7. http://firstmonday.org/issues/issue7_3/dedeke/index.html (accessed March 15, 2002).

Dotinga, R. 2004, July 15. Make a Killing from Antiterrorism. Wired.com. http://www.wired.com/news/business/0,1367,64215,00.html (accessed July 18, 2004).

Dragonetti, J. 2002, May 2002. President's budget request recommends reduced funding for USGS (5/02). *Professional Geologist.* http://www.agiweb.org/gap/legis107/tpg_usgs.html (accessed April 30, 2007).

Dvorak, J. 2004, September 27. Why Online Newspapers Require Registration. http://blog.topix.net/archives/000035.html (accessed January 27, 2005).

EAIS. 2006. Egyptian Antiquities Information System. EAIS. http://eais.org.eg/index.pl/home (accessed February 8, 2006).

Economist. 2001, September 6. Who's Wearing the Trousers? http://www.economist.com/business/displayStory.cfm?Story_ID=770992 (accessed September 8, 2001).

Economist. 2003, November 13. Perishing Publishing. http://www.economist.com/science/displayStory.cfm?story_id=2208619 (accessed November 14, 2003).

Economist. 2005a, October 20. A Market for Ideas. http://www.economist.com/surveys/displaystory.cfm?story_id=5014990 (accessed October 22, 2005).

Economist. 2005b, February 10. Who Pays the Piper ... http://www.economist.com/science/displayStory.cfm?story_id=3644245 (accessed February 11, 2005).

Economist. 2006a, March 30. Battle of Britannica.http://www.economist.com/science/displayStory.cfm?story_id=6739977 (accessed April 1, 2006).

Economist. 2006b, January 19. Extra, Extra. http://www.economist.com/business/displayStory.cfm?story_id=5421892 (accessed January 20, 2006).

Economist. 2006c. March 3. Pumping out Songs — and Pumping up Prices? http://www.economist.com/agenda/displaystory.cfm?story_id=E1_VVJDSJR (accessed March 4, 2006).

EgyMaps. 2006, February. Introducing Egymaps. Quality Standards Information Technology. http://www.egymaps.com/ (accessed February 8, 2006).

Egypt. 1921. *Reports of the Commission on Registration of Title to Land 1917–1921.* Government of Egypt, Cairo.

Elrouby, S., K. Harju, and I. Corker. 2005. *Developing an Automated Cadastral Information System in Egypt.* 10. FIG Working Week 2005 and GSDI-8, Cairo.

Environment. 2003, September. Environment Agency Property Search. Environment Agency, http://www.environment-agency.gov.uk/yourenv/497473/?version=1&lang=_e (accessed September 3, 2003).

Europe. 2006, July 31. Scientific Publications. European Commission. http://ec.europa.eu/research/science-society/page_en.cfm?id=3184 (accessed August 21, 2006).

Evans, P. and T.S. Wurster. 2000. *Blown to Bits: How the New Economics of Information Transform Strategy.* Harvard Business School Press, Boston.

FGDC. 2006. NSDI Cooperative Agreements Program. FGDC. http://www.fgdc.gov/grants (accessed February 14, 2006).

GAO. 2006, June 15. *Information Technology: Agencies and OMB Should Strengthen Processes for Identifying and Overseeing High Risk Projects,* GAO-06-647. GAO. http://www.gao.gov/docsearch/abstract.php?rptno=GAO-06-647 (accessed August 2, 2006).

Geovision. 2002, October 22. Pipeline Mapping in Egypt: A Consultant's Perspective, talk at Glasgow University. Geovision. http://www.geovision.co.uk/Presentations-and-Papers/GU-lecture-Oct02.htm (accessed February 13, 2006).

Goldfarb, J. 2005, October 20. Publishers Become Retailers by Selling Online. Reuters. http://today.reuters.com/news/newsArticle.aspx?type=internetNews&storyID=2005-10-20T151831Z_01_ARM054204_RTRUKOC_0_US-MEDIA-BOOK-FAIR-ONLINE.xml&archived=False (accessed October 22, 2005).

Gomelsky, V. 2006, March 30. Atop Swiss Watchmaking Peaks, Rarefied Air. *International Herald Tribune*. http://www.iht.com/articles/2006/03/30/reports/rwatchtop.php (accessed April 1, 2006).

Graybow, M. 2005, January 7. *New York Times* Mulls Charging Web Readers. Reuters. http://www.reuters.com/newsArticle.jhtml?type=internetNews&storyID=7269738 (accessed January 8, 2005).

Hinsliff, G. 2006, August 6. Brown to Let Shops Share ID Card Data. *Observer* (London). http://observer.guardian.co.uk/uk_news/story/0,,1838315,00.html (accessed August 6, 2006).

Hughes, A. 2001. Content costs and pricing models in the Internet age. *Business Information Review*, 18: 5–10.

Independent (London). 2003, June. Introducing the Independent Portfolio. http://www.independent.co.uk/portfolio/ (accessed June 22, 2003).

Kablenet. 2005, July 27. Patient Phone Charges under Scrutiny. Kable Government Computing. http://www.kablenet.com/kd.nsf/Frontpage/8BDEBC18A26A8F528025704B0048C365?OpenDocument (accessed July 27, 2005).

Kablenet. 2006, September 11. DTI Backs Underground Mapping. Kable Government Computing. http://www.kablenet.com/kd.nsf/Frontpage/48BC223DAD2B37FE802571E30057C18B?OpenDocument (accessed September 11, 2006).

Kintisch, E. 2006, September 6. Putting Patent Trolls on the Defensive. *Technology Review*. http://www.techreview.com/read_article.aspx?id=17459&ch=infotech (accessed September 11, 2006).

Krebs, B. 2001, April 14. New York Times Co. Begins Web Layoffs, Buyouts Next Week. Newsbytes.com. http://www.computeruser.com/news/01/04/14/news1.html (accessed April 14, 2001).

Landmark. 2003a, April. Landmark Brings Court Action against Sitescope to a Successful Conclusion. Landmark Information plc. http://www.landmark-information.co.uk/pr11.htm (accessed March 5, 2003).

Landmark. 2003b, February 21. Useful Information. Landmark Information plc. http://www.envirosearch.info/furtherinfo.htm#contamination (accessed February 21, 2003).

Lash, S. 2002. *Critique of Information*. Sage, London.

Longhorn, R. and M. Blakemore. 2004. Re-visiting the valuing and pricing of digital geographic information. *Journal of Digital Information*, 4: 1–27. http://jodi.tamu.edu/Articles/v04/i02/Longhorn/ (accessed March 27, 2007).

Lufthansa. 2004, November 18. Lufthansa New First Class Terminal in Frankfurt. Luchzak.be. http://www.luchtzak.be/article6515.html (accessed April 20, 2005).

Lulu. 2005, August 14. Lulu Is Free, Fast and Easy. http://www.lulu.com/uk (accessed August 14, 2005).

Mahoney, S. 2006, December 5. Clear Pricing Can Muddy a Message. Mediapost.com. http://publications.mediapost.com/index.cfm?fuseaction=Articles.showArticleHomePage&art_aid=52044 (accessed December 12, 2006).

Masser, I. and M. Blakemore (Eds.). 1991. *Handling Geographic Information: Methodology and Potential Applications*. Longmans, Harlow, Essex.

McCullagh, D. 2005, September 21. NSA Granted Net Location-Tracking Patent. CNET News. http://news.com.com/NSA+granted+Net+location-tracking+patent/2100-7348_3-5875953.html (accessed September 25, 2005).

MCIT. 2005. *Egypt's Information Society*, 4th ed. 61. Communications and Information Technology, Cairo.

Met Office. 2005, December. Aircraft De-Icing Forecast Service. Meteorological Office. http://www.met-office.gov.uk/aviation/deicing.html (accessed December 27, 2005).

Multimap. 2001, October 1. Multimap.Com Awarded US Internet Patent. Multimap.com. http://www.multimap.com/indexes/pressindex.htm (accessed March 23, 2002).

Murphy, C. 2004, August 20. Dear Readers. *Atlantic Monthly.* http://www.theatlantic.com/doc/prem/200409/cullison (accessed August 20, 2004).

Nethouseprices. 2005, February. http://www.nethouseprices.com/ (accessed February 21, 2005).

NFMMC. 2005, July 7. Letter to Office of Management and Budget. NFMMC. http://www.realtor.org/GAPublic.nsf/files/Floodmapltr.pdf/$FILE/Floodmapltr.pdf (accessed August 2, 2006).

NRC. 2003. *Weaving a National Map: Review of the U.S. Geological Survey Concept of The National Map.* National Research Council, National Academies Press, Washington, DC.

Odlyzko, A. 2004. The evolution of price discrimination in transportation and its implications for the Internet. *Review of Network Economics,* 3: 323–346.

OECD. 2006. *Digital Content Strategies and Policies.* 43. OECD, Paris.

OMB. 1990, October 19. Coordination of Surveying, Mapping, and Related Spatial Data Activities. OMB. http://www.whitehouse.gov/omb/circulars/a016/a016.html (accessed June 14, 2002).

OMB. 1992. *Management of Federal Information Sources,* Circular A-130. OMB, Washington, DC.

OMB. 1995. *Electronic Dissemination of Statistical Data.* 98. OMB, Washington, DC.

OMB. 2002, August 19. Circular A-16, revised. OMB. http://www.whitehouse.gov/omb/circulars/a016/a016_rev.html (accessed September 5, 2002).

Openstreetmap. 2006, September. Welcome to Openstreetmap. Openstreetmap.org. http://wiki.openstreetmap.org/index.php/Main_Page (accessed September 12, 2006).

OPSI. 2006, July. Office of Public Sector Information Report on Its Investigation of a Complaint (So 42/8/4): Intelligent Addressing and Ordnance Survey. OPSI. http://www.opsi.gov.uk/advice/psi-regulations/complaints/SO-42-8-4.pdf (accessed July 13, 2006).

Paypal. 2003, March. About Us. Paypal.com. http://www.paypal.com/ (accessed March 8, 2003).

Penenberg, A.L. 2004, August 4. What, Me Register? Wired.com. http://www.wired.com/news/culture/0,1284,64392,00.html (accessed August 6, 2004).

Peterson, B.S. 2006a, November 22. Airlines Shower Perks on Big Spenders. *International Herald Tribune.* http://www.iht.com/articles/2006/11/21/business/loyal.php (accessed November 23, 2006).

Peterson, K. 2006b, February 15. Hotels Corner Market on Online Bookings. Reuters. http://today.reuters.com/news/newsArticle.aspx?type=internetNews&storyID=2006-02-15T183233Z_01_N145123_RTRUKOC_0_US-LEISURE-SUMMIT-ONLINETRAVEL.xml&archived=False (accessed February 16, 2006).

Peterson, K. 2006c, June 9. Internet Travel Agencies Losing Some Luster. Reuters. http://today.reuters.com/news/newsArticle.aspx?type=internetNews&storyID=2006-06-10T022929Z_01_N08385676_RTRUKOC_0_US-LEISURE-TRAVEL-OUTLOOK.xml&archived=False (accessed June 11, 2006).

QSIT. 2005. *Egyptian Geography Network.* 7. QSIT, Cairo.

Rennie, D. 2006, June 17. Eu Gaffe Could Expose the Navy's Chart Secrets. *Daily Telegraph* (London). http://www.telegraph.co.uk/news/main.jhtml?xml=/news/2006/06/17/wmaps17.xml&sSheet=/news/2006/06/17/ixnews.html (accessed June 22, 2006).

Reuters. 2004, December 17. Worker Sacked for Selling Queen's Xmas Pud on Ebay. Reuters. http://www.reuters.com/newsArticle.jhtml?type=internetNews&storyID=7126800 (accessed December 20, 2004).

Robinson, J. 2006, May 28. Mind the Gap: The Press Must Follow Readers Online, but Where's the Cash? *Guardian* (London). http://media.guardian.co.uk/advertising/story/0,,1784605,00.html? (accessed June 5, 2006).

Rose, J. 2001. *The Intellectual Life of the British Working Classes*. Yale University Press, New Haven, CT.

Sayyed Badr, A.-S.A. 1997. Utilities Management in Egypt Governerates: A Unique Experience. GIS Quatar. http://www.gisqatar.org.qa/conf97/links/j1.html (accessed February 8, 2006).

Schenker, J.L. 2006, August 26. Travel Search Engine Gets $10.2m. *Red Herring*. http://www.redherring.com/Article.aspx?a=18269&hed=Travel+Search+Engine+Gets+%2410.2M (accessed August 31, 2006).

Schiff, F. 2003, June. Business Models of News Web Sites: A Survey of Empirical Trends and Expert Opinion. *First Monday*, 8. http://firstmonday.org/issues/issue8_6/schiff/index.html (accessed June 14, 2003).

Schmidt, H. 2003, November 21. Comparing Prices on the Internet Can Bring Savings of 30 Percent. Frankfurter Allgemeine Zeitung. http://www.faz.com/IN/INtemplates/eFAZ/docmain.asp?rub={F040FFD3-897B-46DF-9603-752DD6405389}&doc={722DCB6C-4B8A-4D37-B1C2-94EDB586FD30} (accessed March 4, 2004).

Seelye, K.Q. 2005, March 14. Can Papers End the Free Ride Online? *New York Times*. http://www.nytimes.com/2005/03/14/business/media/14paper.html?adxnnl=1&oref=login&adxnnlx=1110873728-De8Aip4MQFtpMZxUAbLpXA (accessed March 15, 2005).

Shapiro, C. and H.R. Varian. 1999. *Information Rules: A Strategic Guide to the Networked Economy*. Harvard Business School Press, Boston.

Sitescope. 2003, February 21. Neighbourhood Environmental Search. Sitescope plc. http://www.homecheck.co.uk/ (accessed February 21, 2003).

Slater, D., M. Smith, D. Bambauer, U. Gasser, and J. Palfrey. 2005. *Content and Control: Assessing the Impact of Policy Choices on Potential Online Business Models in the Music and Film Industries*. 82. Berkman Center for Internet & Society Research, Cambridge, MA.

Snyder, H. and E. Davenport. 1997. *Costing and Pricing in the Digital Age: A Practical Guides for Information Sciences*. Library Association, London.

Sternstein, A. 2005, October 17. Mapping Technology Threatens USGS Jobs: Fewer Mapmakers Needed as Agency Prepares a Competitive Sourcing Bid. *Federal Computer Week*. http://www.fcw.com/article91117-10-17-05-Print (accessed October 20, 2005).

Survey. 2006a. *Annual Report and Accounts 2005–06*. 76. Ordnance Survey, Southampton.

Survey. 2006b, November 6. Impact of NIMSA Withdrawal on Ordnance Survey. Ordnance Survey. http://www.ordnancesurvey.co.uk/oswebsite/media/statements/nimsa.html (accessed November 10, 2006).

Tamima. 2006. Tamima Group. http://www.tamima.com.eg/ (accessed February 8, 2006).

Taylor, R. 2006, June 30. Check In, Log On, Fork Out. *Guardian* (London). http://technology.guardian.co.uk/businesssense/story/0,,1808590,00.html (accessed June 30, 2006).

Tedeschi, B. 2006, March 29. Comparison Shopping Makes Progress Online. *International Herald Tribune.* http://www.iht.com/articles/2006/03/29/business/ecom.php (accessed April 1, 2006).

Thompson, B. 2006, July 10. Money Makes the Net Go Round. BBC. http://news.bbc.co.uk/1/hi/technology/5164350.stm (accessed July 11, 2006).

USGS. 2004. Geospatial One-Stop: Part I: Capital Asset Plan and Business Case (All Assets). USGS. http://www.doi.gov/foia/BY%2004%20Redacted%20Business%20Cases/GEOSPATIAL_red_cc.doc (accessed August 2, 2006).

USGS. 2005a, October 27. Final Report: The National Map Partnership Project. USGS. http://nationalmap.gov/report/NSGIC_TNM_Report_102705_V6.doc (accessed August 2, 2006).

USGS. 2005b. National Map: Part I: Capital Asset Plan and Business Case (All Assets). USGS. http://www.doi.gov/foia/BY%2004%20Redacted%20Business%20Cases/NATIONAL%20MAP_red_cc.doc (accessed August 2, 2006).

USPTO. 2001a, May 29. Computer System for Identifying Local Resources (Definitions), U.S. Patent 6,240,360. USPTO. http://patft.uspto.gov/netacgi/nph-Parser?Sect1=PTO1&Sect2=HITOFF&d=PALL&p=1&u=/netahtml/srchnum.htm&r=1&f=G&l=50&s1=%276240360%27.WKU.&OS=PN/6240360&RS=PN/6240360 (accessed March 23, 2002).

USPTO. 2001b, May 29. Computer System for Identifying Local Resources (Description), U.S. Patent 6,240,360. USPTO. http://patft.uspto.gov/netacgi/nph-Parser?Sect1=PTO1&Sect2=HITOFF&d=PALL&p=1&u=/netahtml/srchnum.htm&r=1&f=G&l=50&s1=%276240360%27.WKU.&OS=PN/6240360&RS=PN/6240360 (accessed March 23, 2002).

USPTO. 2005a, March 8. Programmatically Calculating Paths from a Spatially-Enabled Database, U.S. Patent 6,865,479. USPTO. http://patft.uspto.gov/netacgi/nph-Parser?Sect1=PTO2&Sect2=HITOFF&p=1&u=/netahtml/search-bool.html&r=8&f=G&l=50&col=AND&d=ptxt&s1='geographic+information'&OS=%22geographic+information%22&RS=%22geographic+information%22 (accessed April 4, 2005).

USPTO. 2005b, March 8. Meeting Location Determination Using Spatio-Semantic Modeling, U.S. Patent 6,865,538. USPTO. http://patft.uspto.gov/netacgi/nph-Parser?Sect1=PTO2&Sect2=HITOFF&p=1&u=/netahtml/search-bool.html&r=42&f=G&l=50&col=AND&d=ptxt&s1=longitude&OS=longitude&RS=longitude (accessed April 4, 2005).

USPTO. 2005c, March 15. System and Method for Geographical Indexing of Images, U.S. Patent 6,868,169. USPTO. http://patft.uspto.gov/netacgi/nph-Parser?Sect1=PTO2&Sect2=HITOFF&p=1&u=/netahtml/search-bool.html&r=34&f=G&l=50&col=AND&d=ptxt&s1=longitude&OS=longitude&RS=longitude (accessed April 4, 2005).

USPTO. 2005d, March 15. Map Data Providing Apparatus, Map Data Installing Terminal Device, and Communication-Type Navigation Apparatus, U.S. Patent 6,868,334. USPTO. http://patft.uspto.gov/netacgi/nph-Parser?Sect1=PTO2&Sect2=HITOFF&p=1&u=/netahtml/search-bool.html&r=28&f=G&l=50&col=AND&d=ptxt&s1=longitude&OS=longitude&RS=longitude (accessed April 4, 2005).

USPTO. 2005e, March 15. Method of Converting Geospatial Database into Compressive Database for Multiple Dimensional Data Storage, U.S. Patent 6,868,421. USPTO. http://patft.uspto.gov/netacgi/nph-Parser?Sect1=PTO2&Sect2=HITO FF&p=1&u=/netahtml/search-bool.html&r=19&f=G&l=50&col=AND&d=ptxt &s1=longitude&OS=longitude&RS=longitude (accessed April 4, 2005).

USPTO. 2005f, March 22. Intelligent Road and Rail Information Systems and Methods, U.S. Patent 6,871,137. USPTO. http://patft.uspto.gov/netacgi/nph-Parser ?Sect1=PTO2&Sect2=HITOFF&p=1&u=/netahtml/search-bool.html&r=12& f=G&l=50&col=AND&d=ptxt&s1=longitude&OS=longitude&RS=longitude (accessed April 4, 2005).

USPTO. 2005g, March 29. Method of Inputting a Destination into a Navigation Device, and Navigation Database, U.S. Patent 6,873,906. USPTO. http://patft.uspto.gov/ netacgi/nph-Parser?Sect1=PTO2&Sect2=HITOFF&p=1&u=/netahtml/search-bool.html&r=2&f=G&l=50&col=AND&d=ptxt&s1=longitude&OS=longitude& RS=longitude (accessed April 4, 2005).

USPTO. 2005h, March 29. Method and System for Controlling Presentation of Information to a User Based on the User's Condition, U.S. Patent 6,874,127. USPTO. http://patft.uspto.gov/netacgi/nph-Parser?Sect1=PTO2&Sect2=HITOFF&p= 1&u=/netahtml/search-bool.html&r=1&f=G&l=50&col=AND&d=ptxt&s1= 'geographic+information'&OS=%22geographic+information%22&RS=%22g eographic+information%22 (accessed April 4, 2005).

Varian, H.R. 1996, August. Differential Pricing and Efficiency. *First Monday*, 1. http:// www.firstmonday.dk/issues/issue2/different/ (accessed May 25, 2000).

Vodafone. 2003. Vodafone Egypt Implements Enterprisewide GIS. *ESRI Telecom Connections*, 8–9.

Walker, L. 2003, October 9. Going, Going, Gone to the Concert. *Washington Post*. http://www.washingtonpost.com/wp-dyn/articles/A63829-2003Oct8.html (accessed October 9, 2003).

Ward, M. 2004, January 21. Snooping Industry Set to Grow. BBC. http://news.bbc. co.uk/1/hi/technology/3414531.stm (accessed January 21, 2004).

Webb, C.L. 2004, April 29. Defense Sector Riding High. *Washington Post*. http://www. washingtonpost.com/wp-dyn/articles/A53032-2004Apr29.html (accessed April 30, 2004).

Webster, B. 2005, November 30. Railways Cut Down on Cheap Tickets to Make Most of Season's Greetings. *Times* (London). http://www.timesonline.co.uk/article/0,,2-1897280,00.html (accessed January 19, 2006).

Wikipedia. 2006, July. Price Discrimination. http://en.wikipedia.org/wiki/Price_discrimination (accessed July 1, 2006).

Willard, A. 2005, November 23. War, Terrorism, Riots Spell Boom for Spy Products. Reuters. http://today.reuters.com/news/newsArticle.aspx?type=internetNew s&storyID=2005-11-23T182943Z_01_FLE356377_RTRUKOC_0_US-SECURITY-FRANCE-GADGETS.xml&archived=False (accessed November 24, 2005).

Wray, R. 2006, June 29. Boost for Free Internet Access to Public Funded Research. *Guardian* (London). http://business.guardian.co.uk/story/0,,1808203,00.html (accessed June 30, 2006).

chapter five

Geographic information, globalization, and society

5.1 Introduction

This chapter explores the nature and role of geographic information* (GI) in contemporary society. Earlier chapters have looked at the value of GI and business and pricing issues, and Chapter 6 will explore the economic and political tensions that impact on the availability of information. This chapter starts by unpacking one of the prevailing myths of GI — that it is everywhere as a fundamental component of all information. It then looks more generally at the politics of information, at the development of spatial data infrastructures, and at privacy and surveillance in the context of GI products that enhance our mobility, but may threaten our privacy. It will examine paradoxes emerging over data protection, data privacy, and anonymity, and the policy-stated benefits of better services to citizens, reduced social and economic exclusion, democracy, and participation, noting key theories about the (geographic) information society.

5.2 The ubiquity of GI

Is GI the most important component of any type of information? It was promoted in the late twentieth century as a fundamental underpinning of the information spaces of government, economy, and society. The often repeated statement is that "around 80% of information is estimated to contain a spatial content" (Lawrence, 2004), an "estimated 80% of government data has spatial component" (FGDC, 2004b), and "Es wird etwa geschätzt, dass 80% aller Entscheidungen eine räumliche Komponente enthalten und durch Geoinformation verbessert werden könnten" (Frank, 2002, p. 11). The 80% claim is replicated without clarification in GI policy from governments (GIPanel, 2005; Scotland, 2006), in a progress report on U.S. presidential initiatives in eGovernment** (OMB, 2006), by industry associations promoting geographic information technologies (GITA, 2006), and by the military (MOD, 2006).

---

* The acronym GI as used in this chapter should be taken as synonymous with terms such as *geospatial information* and *spatial information,* now widely used in much of the literature.
** Fast Fact: Studies indicate that roughly 80% of all government information has a geographic component.

However, it is very difficult to source this estimate back to the original underpinning evidence, although Rob Mahoney (personal communication, May 2005) confirmed to us that he used the figure in evidence provided by British Gas to the U.K. Chorley Enquiry (which reported in 1987; see below), with 60 to 70% of British Gas data being spatially referenced. The figure was later revised to 80% in a presentation at the AM/FM 1988 Conference in Nottingham, U.K., which also marked the creation of the U.K. Association for Geographic Information. In addition, an information audit carried out by Medway Council (U.K.) noted: "Of the 180 database repositories, 121 had some and 11 a possible geographic reference, i.e. around 75% in all. Of the other repositories, 77 or just fewer than 60% had some geographic reference" (Schmid et al., 2003, p. 5).

GI was noted as being a key component of European public sector information (PSI) (PIRA, 2000) and is the subject of a specific European Union (EU) directive, called INSPIRE (Infrastructure for Spatial Information in Europe), which assumed legal force on May 15, 2007, designed to integrate GI within all 27 EU member states. In the U.K., the government review in 1987 (the Chorley Report) argued that GI and geographic information systems (GISs) were as significant for society and the economy as was "the printing press to information dissemination" (Environment, 1987, p. 8). Governments that were not focusing sufficiently on GI were arguably not benefiting the economy and society. In Germany, a study argued that the limited dissemination of GI to the market meant "only approximately 15% of the market volume which could be attained in North Rhine Westphalia has actually been achieved" (Fornefeld and Oefinger, 2001, p. 1). In the U.S., the presidential order establishing the National Spatial Data Infrastructure stated: "Geographic information is critical to promote economic development, improve our stewardship of natural resources, and protect the environment" (Clinton, 1994). Early justification for the European Union's INSPIRE directive focused on GI as critical input to policy development that address the "growing interconnection and complexity of the issues affecting the quality of life today" (Europe, 2004b, p. 2).

One outcome of promoting the centrality of GI was a risk of raising GI and GIS onto a disciplinary pedestal where it could become an easy target for hostile critique. For, as GIS promoted the centrality of information and technology, so geography — the natural host discipline — was in the process of rejecting methodologies that centered on data and quantitative analysis. In the mid-1980s, the quantitative search for order and classification was giving way to qualitative methodologies and the search for difference and uniqueness. While it is too extreme to argue that GI/GIS largely diverged from geography in most geography departments, the quantitative approaches had been a lessening focus in human geography, and mutual critiques often became polarized. Consequently, John Pickles's edited book *Ground Truth* (Pickles, 1995) was an objective attempt to review the prevailing methodology of GIS, but was often taken as anti-GIS. A GIS stores numerical information about reality, such as coordinates and statistical and feature attributes, and therefore imposes a particular digital classification of social, economic, and environmental features of the real analogue world. People are not so much regarded as individuals, but as attributes linked to coordinate space. Roads, paths, and houses are not social spaces where people interact socially and economically, but are assets to be defined as coordinates and to be managed by governments and businesses.

Therefore, as geography explored new concepts of spaces, GIS remained obdurately focused on coordinate space, and 8 years after *Ground Truth*, John Pickles wrote *A History of Spaces*, which eloquently — but in a language that most GIS professionals would find obscure — explored the narrow technological focus of GIS (Pickles, 2003). That is why much interesting research about spatiality has occurred beyond geography, often in sociology. Thus, while the GIS community may map location within physical polygons/areas such as regions, John Urry writes of regions, networks, and fluids, where networks are spatial structures that transcend the physical boundaries demarcated in the GIS, and social spaces act as fluids that may or may not be contained within the polygons: "Fluids account for the unevenness and heterogeneous skills, technologies, interventions and tacit knowledge" (Urry, 2003, p. 42). Fluids are exceptionally difficult to represent in a GIS, which until recently was not good at storing, manipulating, or representing three-dimensional or temporal data, and as human geography moved to embrace sociology, GIS became more isolated from geography.

There were some mediations in the isolation, in what Nadine Schuurman (2000) calls the "factionalisation in geography." She notes that there has been much research on the social impact of GIS, and in its use within participatory societal applications, but these activities are relatively small scale compared to the sales of technologies worldwide. Indicative estimates of the size of the global GIS/geospatial data market vary considerably from $1 billion to $5 billion a year for GIS products, to 10 times that amount for related services and application. Wherever the figure lies in that spectrum, the market is significant, and the role of the GIS vendors in promulgating the technology in developing and developed nations is significant. There is often a tendency to link the technology to the direct solution of societal and economic problems. For example, the Environmental Systems Research Institute (ESRI) argues: "GIS strengthens the welfare of a nation's citizens,"* and the section termed "Democracy and Peace" in its promotional literature claims that GIS can significantly contribute to stable and sustainable development "by helping to inform the public and to allow better access to government."** It is little surprise that critics of GIS can take socioeconomic research and aim to rebut claims that technology has a direct impact on democracy and governance.

---

* http://www.esri.com/getting_started/government/index.html.
** http://www.esri.com/industries/sustainable_dev/business/dem_peace.html.

Thus, a GIS can be used in planning the location of a new hotel (site selection), in identifying the potential customers (geodemographics and drive time), and in assessing risk from environmental events (slope failure and flood prediction). The location aspect of the hotel will allow the data to be used in searches and in Web mapping. The location can be linked then to other data, such as visual tours of the hotel (flash animation, etc.), and the hotel website can link to other geographical information, such as current weather and weather forecasts. That is fine, and it shows the power of GI, but overall what it is showing is the interplay of issues between physical assets and physical events. Let us select a real hotel, the Jordan Valley Marriott Resort & Spa.* It is an excellent hotel for those who wish to visit the Dead Sea, be pampered, and live well. Like most resort hotels it also displays the characteristics of a gated community, where the very clear boundary of the hotel is a border within which guests feel safe, and beyond which is the "local" world of people who generally are only welcome into the hotel space if they either work there or have sufficient resources to consume at the same level as the guests. So while a GIS will show the hotel as being proximate to the local community, it does not easily show the different "spaces" within which the two groups exist — in effect they do not coexist, and therefore the node/arc topology in coordinate terms gives only physical proximity information, not social and economic spaces information. GI and GIS here give only partial information about the local reality, and it is very difficult to use quantitative attribute information to represent the complexities of local spaces.

5.3   Sociotechnical implications of GI and GIS

The main problem with the promotion of the claimed ubiquity of GI, and the role of GI technologies, is that it consequently must be involved with both beneficial and detrimental aspects of technology and society. While there are positive visions, GI also contributes to policy dilemmas about the increasing spatial resolution of GI and the societal concerns over intrusion, privacy, and confidentiality, for example, in the contest over disclosure control (Doyle et al., 2001) in official statistics. The late twentieth century saw a dramatic increase in the resolution and temporal extent of GI, with individual- and household-level data becoming widely produced by both statistical agencies and credit/marketing companies, and with remote sensing devices able to identify and track individuals, e.g., not just satellites, but also sensing, such as CCTV and cell phone tracking. However, it is not a one-way route from good to evil, where a technology developed for peaceable purposes becomes used for hostile purposes.

Military surveillance technologies have been transferred to civilian use, for example, in the Democratic Republic of the Congo, where movement detectors are used to detect the movement of elephant poachers, thus

---

* http://www.marriott.com/property/propertypage/QMDJV.

allowing security authorities to intercept them more effectively (Merali, 2006). The turbulent interplay of the production and consumption of GI and technologies deserves critical consideration. This is not only because there are societal and ethical issues, but also because it provides a useful feedback mechanism for technology producers. It is too easy to dismiss sociotechnical issues, as Michael Blakemore found when presenting these concerns in December 2005 at an international conference in the Netherlands — a GIS vendor representative responded that he did not really see why Mike should present the downsides of GIS, because there were "so many positives about GIS, and we should concentrate on them."

As more information is produced about us as individuals, we may, paradoxically, have less to say in how the information is managed. A dilemma exists in a contest over the production and verification of information — should a citizen be able to see what someone has written about him, and to challenge its veracity? That goes well beyond freedom of information laws, and attaches property rights to information about an individual (Purdam et al., 2004, p. 278). At present, we have some commercial access rights, such as the right to inspect our credit reference information (Experian, 2005), but the integration of health records in the U.K. has shown the general and critical lack of official data property rights, because patients do not have any rights to influence the information written about them by doctors, nor do they have any access rights to verify the information (BBC, 2005b). Perversely, while governments may seem reluctant to allow citizens access to their personal information, businesses often see benefit in allowing access.

In 2006, the U.S. retailer Wal-Mart announced that it would construct a health database for its 100,000 employees, and the employees would be the owners of their data and determine who could access their records (Medford, 2006). Consequential fears do, however, exist in the context of function creep: Would Wal-Mart be tempted at some stage to monitor the records and identify employees who have illnesses that make them less cost-effective? However, only where a citizen has access to his or her health information can any personal management be undertaken, examples being the FollowMe service in the U.S.,* originally established by an individual who needed to have rapid access to the medical records of her son who suffered from hydrocephalus, so that when they traveled, medical specialists could access important information (Economist, 2005a).

It is not surprising, therefore, that concerns about informational identity ownership should lead to contested positions, and this has particularly affected the use and dissemination of official statistics. The global governance of official statistics is provided by the United Nations; it promotes a general mantra that statisticians should aim for "a reasonable balance" between the economic and social benefits of data used, and the need to balance privacy and confidentiality (UNECE, 2001, p. 13). In practice, this balance is very

---

* www.followme.com.

difficult to achieve, and it is easy to polarize views. In discussions over the blurring of information in the U.K. 2001 Census, i.e., intentionally reducing detail so that an individual cannot be identified, one meeting was told starkly of the fear of singular events: "Once a claim of disclosure was made, confidence and trust in ONS would be damaged" (Statistics, 2001, p. 2). So, even the fear of a claim of disclosure was enough to make the U.K. Office of National Statistics reduce detail substantially. It is likely that this disclosure control paradox will become worse in official statistics, as citizens see a policy difference between official and commercial GI producers. It will also be amplified at times where citizens do not trust the channels through which their information is transmitted. In a 2006 survey by the U.S. Inland Revenue Service, 73% of respondents stated that they were fearful about using the Internet for taxation transactions. Three sociotechnical reasons were given: (1) the technology of the Internet was not secure, (2) the methodologies for privacy protection were not robust, and (3) the activity of cybercriminals was high and there was a threat of identity theft (Weigelt, 2006). There are so many paradoxes in the global information society, many of them centering on the need to have instant access to integrated information, which at the same time increases the risk of information loss — and information abuse. It is not just criminals who are a threat, but also those working within the IT businesses. The U.S. Secret Service has assessed the risks of insiders ("current, former, or contract employees of an organization") stealing information (USSS, 2006). The consequence of that is the need for ever more vigilance over the recruitment of staff, and the need to monitor and surveil those staff in their work, for they may be contract employees, hired under uncertain or unknown recruitment policies of the third-party organization. These issues further increase the paradox that our freedom to travel across space leads to more unintended consequences of surveillance.

When providing individual data to a retailer, a customer knowingly opts into the provision of such information, typically indicating acknowledgment of such permission on a form. Official statistics are collected and published by legal mandate, and so providing your data is compulsory in this case. Citizens then have to balance the opt-in and emerging property rights in the commercial sector (see the Wal-Mart example above) and contrast it with compulsion from government, perhaps viewing the latter as increasingly appropriating personal information. Now add in a government desire to integrate information to fight global terrorism (DARPA, 2003; Home, 2004; IPTS, 2003) and citizen concerns over the integration of their data, with GI and GIS being as threatening as it is beneficial. The fuzzy boundary between beneficial use and hostile intrusion is not well addressed in privacy legislation. Curry notes this when assessing the benefits of the move to locational identification in the U.S. 911 emergency response system, thus allowing a much more effective response, with the same technology allowing the potential invasion of public and personal space, i.e., "when the telephone beeps and the ad for Starbucks appears" (Curry et al., 2004, p. 367). Overall, however, the issues relating to the provision and access of personal data can easily paint a picture of government making life difficult informationally, and commerce making it rather easier.

The pros and cons for the utilization of GI and related technologies can be exemplified in the context of health and the workplace. It is surprisingly easy to polarize a debate by identifying only good or bad issues. For example, the positives include:

- Making sure that the patient who is about to be operated on is the person described in the medical records. Avoid misidentification by attaching a radio frequency identification (RFID) chip to each patient and scanning the chip before each action (Kablenet, 2006).
- Remote monitoring of patients who are too infirm to attend a surgery, but whose health problems need regular checking of their condition (Dreaper, 2005).
- Technologies that are elderly-friendly to support e-shopping and access to health services. Active monitoring of the activities of elderly people, particularly ensuring that medication is taken at the prescribed times and in the prescribed dosage, and also checking that their activities are not abnormal (Triggle, 2006).
- Smart fabrics that detect small gestures and signals that may allow quadriplegics to autonomously operate an electronic wheelchair (Singer, 2006).
- Staff using wearable computers in retail distribution depots to speed up the dispatch of goods, reduce waste, and therefore allow lower prices to be charged to customers (Blakemore, 2005).
- The tracking of vehicles and key workers as they travel to check on their personal safety (Anon., 2006).

Some of the cases against would include:

- Pervasive monitoring of elderly people who are in effect imprisoned in their accommodation with only electronic interaction, and with a diminution of privacy and dignity, and a loss of personal autonomy (Abascal, 2003).
- Technologies such as call centers superficially providing egalitarian access to a service, but where the service can use other information (such as caller ID) to link the caller location/identity to geodemographic profiling, and then to prioritize response to the most lucrative or commercially important caller (Bibby, 2006).
- The electronic storage of highly personal details related to health that may be accessed by employers wanting to "scan out" potential employees who have genetic disorders that may result in future health costs to the employer.

- Poor IT security, for example, leading to information on RFID chips being accessed by people who do not have permission to access the information (Boggan, 2006).
- "Is one likely to create a dependence on technologies that is more serious than a dependence on other people?" (Stip, 2005).
- The de-humanization of work and the workplace through humans becoming an extension of the corporate information system (Blakemore, 2005).

It is easy to continue adding to both lists, but there is a risk that the technology producers on one hand, and the social scientists on another, may increase the disciplinary distance between them, rather than explore balances and mediations.

The balance often is identified by engaging critically with the end users, in both the design and consumption of technologies. For example, while remote medical monitoring may enhance medical care while simultaneously diminishing personal dignity, its consumption by many people will be in the context of an often subjective judgment of the benefits and threats. The choice may be: Would you rather have a chip on your toilet seat or a person in the bathroom with you? One of the options allows you to stay in your own home; the other requires you to be in a care environment (Biever, 2004).

5.4   Spatial data infrastructures: governance of GI and public sector information

Even if we accept the myth* that GI underpins most information applications, its governance, production, and distribution can present a paradox. Government agencies, for example, national mapping or cadastral agencies (NMCA) and national statistics agencies (NSA), mostly produce pan-national topographic, cadastral, and thematic information. The transnational governance of the information is then mostly based on nation-state participation, through organizations such as Eurogeographics (European NMCAs), the International Cartographic Association (ICA), Eurostat (European Union statistical information), the United Nations (global statistics and geographic information), and UN agencies such as the UN Economic Commissions for Europe (UNECE, 1992) and Africa (UNECA).

Denise Lievesley worried about the "ecological fallacy" that is generated by a country-level focus, where China has the same data power as Luxembourg, where league lists are generated ranking countries against each other, and where "the need for cross-national data leads to the acceptance of the lowest common denominator" (Lievesley, 2001, p. 15). At a global level, the

---

* That is, myth in the context used by Vincent Mosco, when he wrote about prevailing beliefs about technology: "Myths are not true or false, but are dead or alive" (Mosco, 2004).

integration of GI into spatial data infrastructures (SDIs) is further governed by nation-state-oriented structures such as the Global Spatial Data Infrastructure (GSDI, 2003), Global Map (ISCGM, 2003), and Digital Earth (Earth, 2003). The same scale problems affect these SDIs as affect international statistics, where the cartographic and geographic scale of global SDIs at 1:1 million is their equivalent of the lowest common denominator, and "the institutional de-bordering of global initiatives therefore remains a significant challenge" (Blakemore, 2004). This returns us to the initial observations on the disciplinary distance between GIS and human geography — real-world analogue spaces operate and interact at far more complex levels than the physical borders and areas in a digital GIS representation of those spaces.

SDIs therefore exist awkwardly in the context of generative politics. They are constructed within the political and governance structures of nation-states and transnational organizations, but as Peter Slevin notes, "there is a plurality of sources of authority beyond that of the nation state" (Slevin, 2000, p. 21). Yet another paradox emerges. While nation-states have less and less control over business and global economics, they are building information infrastructures that provide the state with a greater ability to manage its legally-mandated activities, yet also provide information that is of use to global businesses who operate beyond the control of that nation-state. One form of compensation for this lack of control over national space involves recentralizing information control through the availability of funds that are tied to performance metrics that require local government to produce and provide data back to the center (LGA, 2003; ODPM, 2003). Richard Sennett notes this information power contest, characteristic of new public management, observing that while integrated information could empower local government and enable more local autonomy, it is the linkage of policy to resources (and see how this really impacts on geographic information in Chapter 4) that means central government "controls the influence of resources into devolved institutions and monitors performance" (Sennett, 2006, pp. 163–164).

Another approach to maintaining influence and power is to develop uniformity projects. The European Union particularly relies on these, because its executive body, the European Commission, has no direct control over the nation-states that comprise the Union. The Commission's policy is strongly geographically-based, starting with the focus on transnational and interregional policy, leaving internal state policy to the member states under the principle of subsidiarity enshrined in the treaties creating the EU. The EU aims to reduce the economic and social unevenness of Europe, to reproduce Europe as "a more or less homogeneous set of technological zones" where the "densities of technological connections" contribute to economic and social development (Barry, 2001, p. 102). One such uniformity project is the INSPIRE directive (Europe, 2006, 2007) to build integrated access to geographic information in Europe. Like most SDIs, this is a process of infrastructure creation through bureaucracy where "problems of co-ordination, access to information, and power struggles between administrations seem to outweigh the real issue at stake" (Hirschhausen, 1999, p. 429). In Chapter 6, we look at the question of whether the cost to achieve INSPIRE at the European level, or GSDI at the global level, acting through monolithic bureaucracies, is really less than the cost of letting the market operate through the economics of pricing, in the overall cost–benefit assessment of SDI implementation.

In the context of INPSIRE, the European Union acts as what Andrew Barry calls a "regulatory state" (Barry, 2001, p. 26). It acts to transform policy in a classical Weberian bureaucracy of top-down governance. Kanishka Jayasuriya sees this as problematical, noting that the combination of Weberian and Westphalian (assuming definitive boundaries between national and EU policies) governance practiced by the EU, and indeed by most SDIs, is "severely eroded by the structural changes unleashed by globalisation" (Jayasuriya, 2004, p. 498). Jayasuriya proposes "policy capacity" as an alternative framework, the emphasis being on relationships that can deal with the complexities of governance. Using that framework, SDI strategies would set the scene in principle so that a diversity of actors could innovate and develop the infrastructure. Maybe we could envisage "mutating SDIs" that start as particular projects and visions, such as the CORINE environmental data initiative of the 1970s (Rhind et al., 1976), become multiply owned, turn into administrative monsters (Longhorn, 2000), and eventually become liberated to the wider community. Even more critical, however, is the fact that the often esoteric debates on access to information in advanced developed nations mask the very real needs to build both GI and infrastructures in developing nations (Agbaje and Akinyede, 2005; Bassolé, 2005). Paradoxically, the UN — one of the world's biggest bureaucratic monsters — through its Economic Commission for Africa, is providing leadership and coordination in that arena (UNECA, 2005a), while the UN GI Working Group is attempting to implement an organization-wide SDI for UN agencies (UNGIWG, 2007).

Rather than view SDI uniformity projects as linearly developing bureaucratic leviathans, we could also interpret them as initiatives in the context of innovation cycles. One possible framework may be provided by the Perez model of ICT adoption, which sees new paradigms emerging through clusters of innovative activity that attract new and significant areas of investment. Ikka Tuomi evaluates the Perez model in the context of Moore's law of microprocessor development, noting that an initial new paradigm leads to a "gold rush where unrealistic expectations and irrational exuberance dominate" (Tuomi, 2002). "Transient monopolies" are created that can produce significant benefits for investors, but in reality the overall process involves a lot of failure as well as success, and new technoeconomic paradigms arrive with a bubble and crash (Tuomi, 2004). The Perez model may well accommodate colonial interpretations of SDIs, where dominating global GI models (information and technology) are produced primarily by the U.S. GIS industry and the federal information producers who provide significant assistance to SDI development in other nations (Reichardt and Moeller, 2000). Indeed, it is U.S. policy to maintain leadership and influence in global SDI development, and to work with SDI activities in other countries that are "of value to US government, private, and academic interests" (Schaefer and Moeller, 2000, p. 1).

The Gartner Group uses ICT innovation cycles to interpret technological innovation, where early enthusiasm often generates unwarranted expectations, leading to a period of disillusionment. At that stage, an initiative could either collapse and fail, or engage with something like a "killer application" that leads to a "plateau of productivity" when it becomes mainstream (Twist, 2004). The Gartner model would allow us to interpret the current bureaucratic inefficiencies of SDI creation as being at the period of disillusionment, with the killer application for most SDIs being the need to address global warming. Galperin, by contrast, adopts an organizational approach where the ownership of an SDI can influence its success or failure (Galperin, 2004). Ownership can be by a special interest group that builds on common economic interests (p. 160), an ideological approach "through which decision makers interpret complex problems and assess the validity of alternative policies" (p. 161), or a technological approach that is associated with policy and organizational reforms (p. 162). Harmeet Sawhney interprets the ideological approach in the context of physical infrastructure developments, noting that "at the heart of every infrastructure development process is a leap of faith" (Sawhney, 2001, p. 33), where the economic cost–benefit is subservient to the intangible benefits such as political gain. This may explain the previous observation that the EU INSPIRE initiative is not clearly underpinned by a rigorous economic assessment of the relative cost–benefits, although these were attempted (Environment Agency, 2003; Eurostat, 2004), but instead is "crucial to improve environmental policy" (Europe, 2006).

In a later paper Sawhney sees infrastructure development being enacted over eight stages. These stages show a direct contrast to the centralization of SDIs, since the first stage is the "sprouting of islands," and is typified by e-government developments in India, where there is inertia in the creation of an SDI at the central government level, but significant development at the locality level (Hindu, 2005; Umashankar, 2005). In the U.K., regional (subnational) SDIs have been developed in Wales, Scotland, and Northern Ireland, yet not in England or for the U.K. nationally (AGI, 2004). Similarly, in Spain, regional SDI development is well advanced in the province of Catalunya (Guimet, 2004), at both the legal and practical levels, yet much less advanced across the nation as a whole. By stage 5, new infrastructures start to compete with the "old system," which may explain the Egyptian situation outlined in Chapter 3, and in stage 6 they start to subordinate the old system (Sawhney, 2003, p. 27). That interpretation, however, is useful for infrastructures either where there is competition or, as in Egypt, where the private sector creates a new infrastructure because available national mapping is so poor. In many cases, SDIs are more often reformulations of the old structure, rather than replacement of the old structure with a new structure.

More worrying for SDIs, however, is the development of information infrastructures that are beyond the direct control of governments, and which are external to the existing governance of SDIs. For example, Experian, Tele Atlas, Multimap, Landmark, and others are commercial entities that have built significant GI infrastructures, but who are not significantly involved in SDI governance. Overall, these commercial SDIs, and the experience in India, show a centrifugal process forming "emergent structures" (Urry, 2003, p. 29) that are created because the market cannot wait for the bureaucracy to create the SDI. SDI initiatives in Europe and the U.S. in particular are more centripetal processes, where the center generates influence through a process of policy and standards control, and tries to control the creation of the infrastructure. Even this is too simplistic, however, since the centripetal activity of SDI creation is operating at the same time as centrifugal commercial innovation in GI creation and collection.

Reality is more as John Urry sees it — globalized information processes are multidirectional, de-bordered, with "flows of energy, information, and ideas backwards and forwards between the centres and peripheries" (Urry, 2003, p. 83), and all processes interacting with each other. Some political thinking about global policy notes that small, localized but strong political groups may also have an influence on policy well beyond their size and official legitimacy (NIC, 2004). Yet overall, SDIs are rooted strongly to nation-state legitimacy, and within nation-states such as the U.S., there is stronger centralization of decision making into the National Geospatial Programs Office (NGPO, 2005).

SDIs can help promote global governance, and Nelson Mandela was critically aware of the difficulty of controlling national borders, arguing that "it is no longer absolutely certain where countries end, and people begin" (Mandela, 1997, p. 295). During the 1990s there was also the emergence of GI structures that go across the nation-state spatiality of most SDI initiatives, notably the clustering of urban spaces into special interest groups such as the Global Cities Dialogue (GCD, 2003), or the Telecities (2003) initiative that builds on the desire of the European Union to develop cross-border and transnational networks to help create the European knowledge and information society (Dai, 2003). This has led to geographical relationships being partially reprioritized based on similarities across space, e.g., networks of islands, remote rural areas, or geodemographic and cultural or social similarities, rather than the traditional proximity in space. Major cities form transnational structures, since there is the possibility of "the dislocation of the city, its overextension and disappearance" (Crang, 2000, p. 301), where the relationship of a city may be stronger with other cities rather than its geographical hinterland, or where cities such as London, Los Angeles, or Tokyo are so large that they do not operate as an entity.

Paradoxically, therefore, GI increasingly allows "action at a distance" and contributes to the dilution of locality. The integration of GI into the infrastructures further enables global capitalism to neglect, or bypass, the "remaining portions of national territories" that are not profitable or productive, thus undermining the "relatively standardized and equitable infrastructure systems" of the post-Second World War Fordist and Keynesian political and economic systems (Graham and Guy, 2003, p. 379). Historically, social assets such as water and electricity, operated and owned from the public sector, have increasingly become privatized, where sophisticated GI and GIS underpin the marketization of essential services such as water (Lievesley, 2001, p. 4). Joseph Stiglitz warns that the operation of core social utilities as capital markets "is inevitably accompanied by huge volatility, and this volatility impedes growth and increases poverty" (Stiglitz, 2002). There then emerges an almost circular paradox that GI is embedded into information infrastructures that aim to overcome (OECD, 1996) the social and economic exclusions (such as the generic digital divide) that the availability and use of GI has unwittingly helped to develop, for example, through spatial customer segmentation.

5.5   GI globalization: mobility, location, and boundaries

The preceding discussion underlines the characteristic production of much GI being strongly rooted in national governments and their institutions. Residing in fixed-location information systems such as GIS, GI then empowers mobility. John Urry (2003) develops complexity theory to help argue that the twenty-first century "will be the century of inhabited machines" (p. 127) that form the "moorings" that enable the "mobilities" of globalization (p. 138). It is the interplay of the machines, inhabited with such things as GI and software, that facilitates our abilities to travel, interact, and undertake business across time and space (Urry, 2003, p. 126). The moorings then become nodes on the interconnections facilitated by the Internet, with its openness and accessibility, but also with its "placelessness" that makes it so easy for people to interact across space, and to avoid the traditional legal, ethical, and moral constraints of place-based interaction (Naughton, 1999, p. 269).

Martin Dodge and Rob Kitchen provide a different perspective on moorings and mobilities through their analysis of code–space. Code–space is constructed through the classifications (computer code that classifies data) of credit reference and geodemographics information systems. Through the classification of census and our spending (credit and charge card) information, spaces are created that identify groups such as high spenders, impoverished communities, etc. They argue that "the code exists in order to produce space" (Dodge and Kitchin, 2004, p. 209). They note a dyadic relationship between code and space, since space is encoded through coordinates and attributes, and in the moorings of a GIS a new space is produced and managed. Stephen Graham delves deeper into this dyadic relationship, noting that the systems that we use often are black boxes, where we understand little about the proprietary algorithms and models that process our data and produce results (Graham and Wood, 2003). That makes it very difficult for us as individuals to challenge the classifications, since even if we are experts in spatial classification, the algorithms that are used in the code systems often are proprietary information.

Moorings can themselves be threatened by increasing mobilities. A major example for the European Union is the problem of policing borders. As the EU has expanded, and has created a larger internal space of mobility, the farther borders of the EU have become porous, and illegal immigration has increased. The EU Borders Agency (Eupolitix, 2003) was therefore established, hosted by Poland (Kubosova, 2005), which is one of the states that has part of the outer EU border. This then links to the "political rescue of distance" (Robins and Webster, 1999, p. 249). The border agency will make extensive use of GI and information relating to the identity of citizens; for example, biometrics, integrated information, information sharing, and secure technologies feature in this initiative. In order that the integration of data for border surveillance is not seen as a Big Brother activity, there is an associated political initiative to persuade citizens that their privacy will not be eroded. Indeed, the European Union argues that our privacy could be enhanced, because "they are able to authenticate a person's access rights" (Europe, 2005). What they mean is that as we move rapidly through physical space, e.g., traveling, crossing borders, purchasing goods in shops, checking into hotels, etc., we want to quickly establish that we are who we are, and that we can instantly spend money. At this stage, GI becomes embroiled in the contest between positive and negative outcomes for society in the context of "dimensions of unintended consequences" (Lash, 2002, p. 50). There is a long history of this occurring in technology, for example, the introduction of the automobile, which generated increasing pollution and started the process of depleting critical fossil fuels (Rivers, 2002).

Gary Marx is strongly critical of the rhetoric of arguments that the more we integrate information, the more we are protected in globalization and mobility. He provides a list of "information age fallacies" (Marx, 2003). He contests arguments that more investment in more data and more technologies leads to linear positive outcomes. In particular, he confronts the political rhetoric that is used to challenge terrorism. John Ashcroft, former U.S. attorney general, following the 9/11 attacks, argued in favor of more information about citizens being collected on the basis that "we're not sacrificing civil liberties. We're securing civil liberties" (Crampton, 2003). Crampton notes that this implied that our rights to privacy are always circumscribed. In the U.K., fear of crime is used to capitalize on a willingness to be increasingly monitored by CCTV in public spaces (Fussey, 2004). Nevertheless, Gary Marx stresses the iniquity of the fallacy that states "if you have done nothing wrong, you have nothing to hide" (Marx, 2003, p. 28).

Sewell and Barker are stronger in their criticism of the call that we "subjugate ourselves to surveillance" (Sewell and Barker, 2001, p. 195), noting that surveillance is at the same time both positive and negative for us. The "actuarial and managerialist" culture of administrations (Fitzpatrick, 2002, p. 373), characterized by the collection and monitoring of information about citizens, imposed further erosion of individual privacy because more faith is placed in the information systems than is placed in the citizens to whom the information refers — hence the difficulties noted above of citizens having access to their own data. Indeed, Haggerty and Ericson see the collection of information into a "surveillant assemblage" marking the "disappearance of disappearance" (Haggerty and Ericson, 2000, p. 619), where not wanting to be seen is taken as implicit evidence that we are guilty of something. These systems then not only allow us to be included, for example, identified as legitimately within the borders of the EU, but also can create new social and economic exclusions both in public spaces and in cyberspace (Wakefield, 2004).

The flexibility of GI in helping to enable the mobilities of globalization has been provided not just by the nature of the data, e.g., rapidly developing coverage, resolution, and timeliness, but also by the ways in which GI has been made available through costing and dissemination models (Craglia and Blakemore, 2004; Longhorn and Blakemore, 2004). The latter part of the twentieth century saw a rapidly emerging process of repurposing GI by actors who were outside of the traditional government, or official, users. From the 1960s onward, Census of Population (Census) statistics in the U.S. were used by the commercial sector to classify areas into informationally homogenized marketing zones. Geodemographics rapidly emerged to underpin target marketing, customer tracking, and credit referencing. Indeed, the massive moorings of computer and telecommunication systems, such as those run by Experian (2004), are central to our ability to move seamlessly and fluidly through global space and use plastic money to consume products and services. On that basis, it could be argued that the best GI infrastructures (SDIs) are built beyond or outside government, using existing and emerging global standards and information and communication technology (ICT) infrastructures, yet paradoxically most SDIs have been constructed under government-oriented structures.

With geodemographics, "the complexity of life is reduced to abstract information that permits the construction of a programmed, mediated reality of tastes, behaviours, values and lately experiences" (Arvidsson, 2004, p. 466). Through these systems of classification we no longer are individuals, but are part of a consuming tribe. The increasing collection and storage of GI-related information about our lifestyles externalizes our memory into moorings that are owned by others. Blanchette and Johnson critique the "relationship between social forgetfulness and information technologies" (Blanchette and Johnson, 2002, p. 43), noting that the power is shifting from personal memory to institutional memory, where the externalization of our memory into geodemographics and government databases means that while we may not remember, the information systems never forget. Therefore, GI is both representational of reality and central to the many artificially constructed realities of globalization.

Citizens are classified using cluster analysis in the context of e-government services in the U.K. as e-amenable progressives, contenteds, disenchanted, skeptics, dissatisfied traditionalists, and left-behind traditionalists (MORI,

2004). The classification of individual citizens is mirrored by the classification of the financial health of businesses and organizations by companies such as Moody's, Standard & Poor, and Fitch, that form a natural oligopoly of organizations making life-and-death statements about organizations, and which operate in a market that "is curiously devoid of competition and oversight" (Economist, 2005b). It is in these contexts that the widespread availability of GI is used in software-based exclusions of people and organizations from society and economy. For example, if you do not have a bank account and a credit card, and thus are less able to be classifiable geodemographically, you are significantly less able to participate in global consumerism.

5.6   Repurposing of GI: benefits and risks

The repurposing of GI has been affected by two further processes: time, i.e., the acceleration of processes across space, and an increasing sophistication of repurposing through what Scott Lash (2002) terms "stretched productive relations." This has extended the GI supply chain beyond that of owning and using data, to a sophisticated and demanding dependent relationship where it is increasingly difficult for GI producers to understand the extent of the repurposing of their data, yet where the diverse users place more demands on data producers to provide a sophisticated supply chain with new data and refined existing data. The demands exist because of the sophistication of the GI market, which goes well beyond the "pouring a familiar content into another media form" (Bolter and Grusin, 1999, p. 68) to the production of new types of data and applications. For example, the U.K. Meteorological Service reduced errors in its weather forecasting by 11% when it introduced a new supercomputer and a refined forecasting model (Kablenet, 2005). Hence, GI producers are regrouping the dispersed demand within contractual relationships such as licensing and value-added reseller contracts (Longhorn and Blakemore, 2004) so that they can remain close to user needs. If the GI market is to change from supply driven to demand driven, then it is imperative — and difficult — to better understand just what the demand is for ever more diverse types of GI arising from an ever more diverse user base.

There is, however, no linear relationship between the volume of information available and quality of use, as witnessed with the problems caused through information overload (Shenk, 1997). The GI organizational capacity of agencies to process information may not meet the time imperative imposed by events. This was starkly evident in the U.S. intelligence agencies prior to 9/11, with the congressional investigation noting that the U.S. government had "a weak system for processing and using" its information (Congress, 2004, p. 417). This subsequently generated interest not in the regularities and predictabilities of the information landscape, but in unevenness and unpredictability, one example being Atypical Signal Analysis and Processing (ASAP) (Hollywood et al., 2004). It is a fundamental tenet of SDIs that they need to be in place so that environmental unpredictability can be assessed effectively, and the modeling of catastrophic events such as the 2004 Indian Ocean tsunami is a case in point (ENSI, 2005). Nevertheless, there is not a simple linear relationship between GIS and wider societal benefits, in spite of statements such as "GIS will evolve into a kind of nervous system for our planet" (Dangermond, 2001). This uneven relationship is characterized by Joseph Stiglitz's "information imperfection" thesis where concern is not just about uneven information production, but also uneven access to the technologies, skills, and tools to use the information (UNECA, 2005b). A PEW study into the Internet further advised strongly against "technological determinism," since many changes are "spurred by multiple forces," and where "many were sceptical about advances outside their areas of expertise and were enthusiastic about those in their areas of specialization" (Fox et al., 2005, pp. 47–48).

The more there is a need for faster decision making, often promised by embedding GI into new technologies, the more will be the risk that errors will be made, such as in the area of biometrics and border control, where the European Commission (Europe, 2005) warns that decision makers should take critically realistic viewpoints about the benefits and risks of such technologies. Perhaps here we will see the rise in collateral GI damage through its reuse beyond the original collection purposes, a process sometimes called de-purposing. Here the damage caused to a citizen may be balanced against the greater societal need, or existing access rights to GI and its channels of dissemination are damaged as a result of global terrorism and governmental reactions to terrorism (Defense, 2004; Reuters, 2002). Problems through de-purposing also arise through the inability of an existing dominant GI product to remain strategically ahead of emerging competing products. This has been most evident with the Census of Population in the U.K., where local government now is able to produce more accurate (which really means less inaccurate) data than central government. This introduces yet another paradox, and it is one that challenges SDIs. National or pan-national data collection aims to enable comparison through harmonization, yet harmonization to date always dilutes thematic, temporal, and spatial resolution. The only protection for this loss of detail has been the official label, and the difficulty for other agencies to successfully contest the quality of the official data.

The contest with the role and authority of the Census of Population is important because it can underpin the allocation of electoral representation and can also be tied to resource allocation by central government, where allocative and authoritative resources are central to government control (Robins and Webster, 1999, p. 92). Therefore, following the U.S. 2000 Census, challenges occurred from localities that were concerned about undercounts leading to loss of congressional representation (Smith and Stewart, 2003) and loss of tax revenues (Lavan, 2003), resulting in federal recommendations for increases in Census quality for 2010 (GAO, 2004). In the U.K. the contest was at the city level, with Westminster (London) and Manchester particularly challenging the official statistics on the basis of their own surveys (Statistics,

2003a, 2003b). This contest is lose–lose for central government, since the challenge will arise only in the event of a locality losing something as the result of central GI. As Professor David Rhind, the chair of the U.K. Statistics Commission, said to the government enquiry into the Census, "I know of no local authorities which have complained because they have got more money" (Commons, 2003). To some extent, the increasing observation of those who govern is that the "panopticon has given way to the 'synopticon' where the many are watching the few" (Bauman, 2000, p. 85), as well as the few (government) surveilling the many (the citizens). What Bauman means is that the authority of central government PSI is increasingly being challenged on the basis of evidence, for example, using more accurate local data, rather than judgmental views. Central government data may have power through the allocative mechanisms of finance, but they increasingly lack the trust of those who are at the receiving end of decisions, or who are reusing the data.

For the individual citizen/consumer, debates such as those concerning the Census may seem distant, but there are individual contests that are deeply embedded in both the at-a-distance lifestyles in developed nations, and the at-a-distance supply chains of information, products, and services that are consumed. Life to a large extent is "metricated" through interconnected information and systems, with GI deeply embedded in the metrication. We mentioned the concept of code–space earlier, but there are more practical applications as well. For producers of food, GPS and GI enable wine makers to closely monitor crop development and to micromanage the vineyard planting strategy (AP, 2004). The interconnected supply systems of global supermarket chains stretch their productive relations, e.g., sourcing material from around the world, while also increasing their control over the livelihoods of workers in distant countries, who, in spite of attempts to deliver more information to them, are ever more unable to compete effectively with the global agricultural businesses (Bakyawa, 2005). Global transportation and logistics companies quickly deliver products to outlets, ideally breaking down historically linear supply chains and enabling networked supply chains "to meet the market's wild demand swings" (Forrester, 2000).

In effect what we are seeing is a *de facto*, albeit uneven, food information infrastructure. It is one that is emerging piecemeal out of business strategy and the reactive intervention of governments. In the absence of the moorings of integrated information systems, animals can be transported large distances to markets with few systems in place to monitor the movements and model the possible risks. This mobility of animals within modern industrialized agriculture has led to catastrophic breakdown of quality through foot-and-mouth outbreaks in the U.K., mad cow disease and its human variant, and SARS (severe acute respiratory syndrome), which threatened global capital by traveling along the vectors of international travel. Richard Sennett (2006) writes of the uneven consumption of public resources that occurs with such events. The SARS outbreaks in 2003 killed relatively few people, whereas malaria kills thousands a day, but global and national agencies invested significant resources in attacking the problem of SARS. To be fair, there were many pressures, ranging from heightened public fear of SARS and a reduction in travel, therefore affecting global business viability, and the unknown risk that SARS could quickly spread to a global pandemic. As the *Economist* wrote, however, "simple cheap public health measures (condoms, lifestyle, quarantine) usually work much better than expensive technological fixes like vaccines," (Economist, 2003b) but the pressure to deliver a high-tech fix is very strong — as is the pressure often to use a GIS to justify, underpin, or deliver a policy fix.

Such a policy fix can be a proposal to create new moorings of information, such as the European Union proposal following U.K. foot-and-mouth "to set up IT systems to track every livestock animal in Europe" (Kablenet, 2003). The locational technologies that also empower farmers (GPS, satellite imagery, GIS, etc.) become technologies of surveillance when the activities of farmers are monitored remotely to check whether they are conforming to regulations (Elliott, 2005). Then the extended food chain itself becomes exposed to hostile intervention through agroterrorism. In a review of weaknesses in the response capability of the U.S. agriculture system, the Government Accountability Office highlighted the fact that while information existed, "there are weaknesses regarding the flow of critical information among key stakeholders" (GAO, 2005, p. 7). There is a general level of pessimism about the extent to which the U.S. information and physical infrastructure can be protected from global terrorism, with 66% of experts expecting that "one devastating attack will occur in the next 10 years" (PEW, 2005a).

In the U.K., attempts to build information to prevent another BSE (bovine spongiform encephalopathy) outbreak met problems because they were "inefficient, overly burdensome and based on obsolete technology" (Kablenet, 2004). Yet again, Urry's moorings had been absent, in spite of the fact that there are extensive mobilities in the agricultural system. The failure to create the moorings, however, is less the fault of data availability and more the fault of organizational failure to use the information effectively (Koontz, 2004). There may also be duplicated moorings because "too often agencies are buying the same data and same applications over and over" (Miller, 2003). This situation exists in the U.S. in spite of a 1990 initiative to improve coordination of GI activities in the federal government (OMB, 1990), and that initiative was a revision of a policy of data integration and sharing that dates back to 1967.

What these examples indicate is that there is a clear macroeconomic case for the widest availability of GI, so that moorings can be created to enable the mobilities, or at least improve where the GI is available so that the uncertainties of the mobilities can be monitored.* This takes the debate to

---

* In SDIs such as the European Union's INSPIRE, environmental uncertainties are well acknowledged, such as extreme weather, earthquakes, landslides, etc. However, social and economic uncertainties are not as well covered, because the legal mandate is environmental.

a level higher than "we will save this amount of money," and GI becomes a strategic benefit that must be delivered in response to the turbulent processes of globalization. Moorings are also developed for citizens who are consuming the results of mobilities — the transparency of a food chain being one example, where trust moves beyond a "mark" that says someone else has accredited the product, to "a relationship based on the sharing of information" about the supply chain (Tapscott, 2004). There are both pushes and pulls for this increased openness. The push is the destabilizing of consumer trust, driven particularly by the difficulty of metricating trust at a distance when using e-commerce services, e.g., from e-tailers to auction sites to extra-territorial suppliers of medicines. The pull is the intervention (remediation) of governments who have a responsibility to citizens, and social and professional organizations that have a responsibility to groups of citizens.

For example, there are trust sites for U.K. Muslims who need assurance that meat is really Halal (Boyd, 2004). Such mediations are increasingly important, because the public trust less and less the conventional mediations of the mass media. Indeed, in the U.S. "the public has lost more confidence in the media than in any other major institution in American society" (PEW, 2005b).

5.7   *Information overload, emergent societal spaces, and modernity*

Yet another paradox is emerging through the availability of GI, which is the extent to which the volume of information is creating "noisescapes" that citizens find difficult to navigate through. This in large part is due not just to the volume or overload of information, but to the complexity of debates to which citizens are exposed. For example, Bulkeley claims that "climate change" is an artificial social, economic, and political construction based not only on data, which is not evidence until interpreted, but also on multiple interpretations in the context of discourses of what are the risks to the environment and society (Bulkeley, 2001). Noisescapes also are a product of the "action at a distance" (Slevin, 2000, p. 17) that exists in contemporary society. Information landscapes in preindustrial societies were more homogeneous, more held in the memory of the residents, and less reliant on technological aids. Indeed, the strong identity with a locality and community was one way to maintain citizen conformance in the past, while the concept of banishment from your locality (Kingston, 2005) is less feasible in contemporary globalized societies.

Contemporary society also relies increasingly on artificially represented spaces, since so many of the spaces we function within are disparate and unknown to us, and we need aids that help us "to overcome the tyranny of space" (Hine et al., 2000, p. 1768). Hence the development of vehicle navigation systems that use the calculable space enabled by GI, with initial developments that were expensive and exclusive now being low cost, real time, and ubiquitous. The result of the technology becoming cheap enough for mass consumption is that there will always be uncertain outcomes through what is seen by some as technological failure, and by others as a naïve overreliance on the technology by users. Imperfections in vehicle navigation systems certainly caused problems in the U.K. when people were directed up farm tracks, or even over cliffs; and when too many people used real-time systems to avoid traffic jams, the number of them who were sent on the same diversion route simply caused a new traffic jam elsewhere (BBC, 2006a, 2006b).

That means that the innovation cycles need to keep differentiating the navigation systems to maintain premium offerings that attract the highest prices, or where what we regard as public space when driving is re-regulated into smart highway systems that influence the speed of each car so that highway capacity is maximized (AP, 2005b). Andrew Barry provides an extensive critique of the technologies that are used to evaluate calculable space, questioning also how we can realistically define safe levels for pollutants across space when the measurement of pollution usually is highly generalized from a very few point samples and at limited times (Barry, 2001, pp. 169–170).

The provision of increasing levels of environmental information also introduces complex feedback effects. Not only can citizens see the impact on the environment of global warming and carbon emissions, but they "can also be held to account if new information systems make the environmental impacts of individual consumer choices transparent" (Berkhout and Hertin, 2004). There now is a complex ballet of information production and consumption. For routine, often quite low cost purchases, consumers are presented with a wide variety of cost/price comparison websites (Schmidt, 2003), and some sites will use a consumer's purchase history to target price comparisons (Hill, 2003). It seems rather strange that routine purchases elicit such attention, but it is the routines of life that generate the body of data allowing retailers to profile a customer (AP, 2005a). Conversely, a house purchase, which usually is the single biggest investment a citizen makes, historically has involved very little informational reinforcement by the citizen, with most of the responsibility externalized onto an intermediary such as an estate agent. The U.K. has experienced the emergence in particular of services that empower customers with information, usually focusing on risks such as being on polluted land (Landmark, 2003) or the possibility of flooding (Environment, 2003).

However, the feedback effects are to a large extent circular. While more information is accessible to customers/citizens, even more information about those customers and citizens and their localities is available to business and government. This enables such services as the individual targeting of risk for insurance (Norwich, 2004), which will both reduce costs for locations not at risk and increase costs for those that are potentially* at risk.

---

* And here we enter the ethical contest over the prediction of risk. To what extent can citizens challenge the risk, and will the forecasting models used by the businesses/ agencies be placed in the public domain so that they can be critically evaluated. There is a potential for a Marxist critique of the availability of detailed GI.

The Environment Agency service noted above has led to fears that houses in at-risk areas "could see their values plummet" (ITV, 2004). This leads to the need for increasingly discriminating data, and the GB Ordnance Survey announced in March 2005 that it was building Land-Form PROFILE Plus at a resolution of 50 cm, including elevation — not provided in existing services (Cross, 2005). The very democracy of information availability itself generates new forms of social and economic exclusion, and those in turn generate new market opportunities for refined products.

While much of the above material covers the "new," most of the new is deeply rooted in historical contexts. The ancestry of most official GI exists long before the space–time distanciation of globalization, i.e., "the re-ordering of time and space facilitated by action at a distance" (Slevin, 2000, p. 200); indeed, it existed well before industrialization. The Ordnance Survey of Great Britain (OSGB) was established in 1791 as a military response to poor mapping in Scotland and the disruption this caused to military mobility. OSGB was central to what Bauman terms "heavy modernity," where the manufacturing industry built products on the basis of strict design and production control (Bauman, 2000, p. 47). Warfare was a major stimulus to the collection of GI, and World War II enhanced the role of the geographers and set the structure for post-WWII geographic research (Clout and Gosme, 2003). In post-Second World War Europe, geographic information contributed to the heavy modernity planning systems, where the physical urban environment was analyzed, modeled, and topographically demarcated. Urban information systems were to some extent a precursor to GIS, but both systems needed hard, quantitative, mathematical renditions of space, and those renditions were used in rational and scientific planning approaches.

The transition to "liquid modernity" (Bauman, 2000) in the late twentieth century involved much more uncertainty and much more rapid innovation rates.Bauman cites Nigel Thrift's "soft capitalism" — an economy of business and marketplace disorganization where business and organizations can only respond strategically to the disorganization by being ever more in control of information and its analysis (Thrift, 1997). Thus, GI is a vital locational component of the strategic response, but it is still produced and disseminated by organizations that are grounded historically in heavy modernity. It is no surprise, therefore, that there are so many tensions over the organization metamorphosis of national mapping agencies (NMAs) such as the Ordnance Survey from military structures to business-oriented trading structures that are expected to produce an operating "profit" for government (Survey, 2005). These changes are influencing the ways in which NMAs around the world structure themselves; for example, in India (Nag, 2002), where the military structure of the agency has for a long time dominated its behavior.

A critical constituent of heavy modernity, and one that helped to fuel both manufacturing and information production, was the warfare that stimulated most of the century's investment in GI (Barry, 2001, p. 44). Military priorities such as the Cold War were "at the heart of the information revolution"

(Robins and Webster, 1999, p. 159), with the need to develop spatial data models for cruise missiles (Richardson, 1977), tactical battlefield systems, and satellite imagery and monitoring (Ball and Babbage, 1989). The military priorities, particularly in the post-WWII Cold War period, generated significant investment in the production of GI and the technologies of GIS and remote sensing. This was broken only by a short "peace dividend" in the 1990s (Coghlan, 1994) following the collapse of communism, which lasted with some fragility through to the events of 9/11, and which was assessed in 1998 by ESRI in promoting GI to "support situational awareness" in the reemerging conventional battlefield (ESRI, 1998, p. 3). By 2003, the dividend was long since exhausted and exacerbated by aging populations and a declining base of direct taxpayers in many developed economies (Economist, 2003a). The peace dividend transformed again into the military dividend (Europe, 2004a), with global terrorism leading to demands for more data (Roberts, 2004), better technologies, and some suppression of data previously available to the market (Defense, 2004).

Theorists about globalization place great emphasis on the changing roles of time and space. Antony Giddens in particular uses time–space distanciation (Wikipedia, 2006). Ash Amin stresses the changing relationship of space, place, and time, with these moving away from a Cartesian system, the coordinate base so fundamental to GIS, to a relational organization. Places are no longer the sum of the practices that are contained within them, "and what happens in them is more than the sum of localised practices and powers, and actions at other 'spatial scales'" (Amin, 2002, p. 395). However, was heavy modernity the only time when information flows were very physical? Probably not, and in the history of GI, the framework that may best be used is flow-enhanced disintermediation, wherein "embedded old intermediaries are displaced by disembedded new intermediaries" (Lash, 2002, p. 207). Consider the Internet airline booking business, which started first with airlines providing online booking, then the growth of intermediators such as Expedia.com, followed by strategic remediation by airlines (Opodo.com). The next stage was for airlines to encourage customers to "stick" to their sites by providing the best offers only on that site, and the low-fare airlines added complexity by only allowing booking through their sites, until another intermediator was created (Openjet.com). Overall, as Evans and Wurster note, "disintermediation used to be about substituting reach for richness. Now it is about transforming both, often simultaneously" (Evans and Wurster, 2000, p. 97).

But the process of flow enhancement may not be that new. The printing press generated a flow enhancement that enabled new intermediaries (not just the church) to disseminate information in the Renaissance. It destroyed the clerical monopoly on information and knowledge, and like the modern Internet, it opened access to the general population (Rose, 2001, p. 13). The development of libraries in nineteenth-century Welsh villages allowed miners to be strategic about "enduring prolonged structural unemployment"

(Rose, 2001, p. 251), rather than just idly waiting for work to resume. The English *Domesday Book* of 1086 was a new form of remediation by William the Conqueror, who, through integrated information collection and storage, centralized control over the cadastral landscape of his kingdom. The world's largest current infrastructure, which has emerged without the structured form of coordination practiced by SDIs, is the telephone system. While there is a global form of governance through the International Telecommunications Union (ITU), the current global telephone system interconnects the newest GPS mobile phones with the oldest landline devices in developing countries,* and it does this through *de facto* as well as *de jure* processes.

The electronic telegraph in the nineteenth century enabled new forms of mediation and informational control. Tom Standage's wonderful book *The Victorian Internet* shows strategic disintermediation over price control taking place in Aberdeen, where fishermen could notify markets of what they had caught, receive information about prices elsewhere, and receive orders for their products as well as being much more aware of market conditions (Standage, 1998, p. 159). Disintermediation was initiated by the electronic telegraph when the British government in the 1850s had to stop giving sensitive military information to the *Times* newspaper. Before the telegraph was invented, the newspaper would publish information about military intentions, with all parties knowing that it would take too much time for the enemy to physically transport the information to their governments. Once the electronic telegraph allowed the enemy to transmit the information rapidly, the *Times* and its readers were cut out of the informational loop (Standage, 1998, p. 145), leading to public anger and distrust toward the government. Is that dramatically different from the data-scrubbing post-9/11 (FGDC, 2004a)? New forms of business organization were enabled by the telegraph, notably "the rise of large companies centrally controlled from a head office" — another strategic remediation enabled by informational flow enhancement (Standage, 1998, p. 197). Perhaps, as Jonathan Rose warns when researching literacy history, the history of GI "has been written mainly from the perspective of the suppliers rather than the consumers" (Rose, 2001, p. 256).

5.8 GI consumption: technology and property rights issues

Consumption of GI is performed using the tools and techniques that together comprise the technologies of GI, and it is here that there is a problematical situation early in the twenty-first century. Just as information is becoming more readily accessible, many familiar, even common knowledge techniques are becoming less accessible through the privatization of knowledge via the patent system. Multimap has patented the technique of clicking on

---

* We are grateful to Robert Barr for this observation.

a displayed map to obtain information about the location (Multimap, 2001; USPTO, 2001a, 2001b). A quick search of the U.S. Patent Office decisions in March 2005 shows that patents are increasingly being granted to techniques that previously may have been regarded by the wider community as common knowledge (USPTO, 2005a, 2005b, 2005c, 2005d, 2005e, 2005f, 2005g, 2005h). Just as Harlan Onsrud (1998) writes critically about the need to preserve the information commons, there is a similar need to preserve the knowledge commons by reforming the now overloaded patent systems (Marks, 2005).

One possible defense of such behavior is the severe impact of time–space distanciation on intellectual property, with extensive cyber crime in the context of theft of intellectual property rights (IPR) from all sectors. As the extent of IPR theft has become apparent, so the laws have become more restrictive. The Gartner Group argues that the music industry is "the first to face the potential benefits and terrors of digital distribution" (GartnerG2, 2005, p. 52) — a point that can be clearly contested by NMAs such as the Ordnance Survey GB, which has for a long time been protecting digital IPR (Survey, 2001, 2004) and pursuing those who breach copyright rules. As David Rhind noted, NMAs who protect their copyright aim to persuade users that "information can be a commodity owned by someone else and unauthorised use of it is tantamount to theft" (Rhind, 1996, p. 11). However, the history of GI provides many examples of IPR theft that led to significant innovations. The medieval portolan charts were constructed from information that was gleaned from other sources, and the *Theatrum Orbis Terrarum* of Ortelius in the 1570s at least acknowledged sources such as Mercator and Saxton, but there was no formal exchange of royalties for the use of their IPR.

More recently, an analysis of IPR use by the emerging U.S. economy in the 1900s shows that there was cavalier disregard for the IPR of Europeans. The U.S. government gave patent rights to artisans who brought innovations to the U.S., and indeed offered financial incentives if they arrived with innovations (Ben-Atar, 2004). The only question regarding patent rights was whether anyone in the U.S. had already patented the innovation; it was of no consequence if the artisans had stolen the designs before leaving Europe. Ben-Atar argues that the rapid growth of the U.S. economy was significantly assisted by IPR theft. Yet, the U.S. is at the vanguard of World Intellectual Property Organization (WIPO) calls for the aggressive protection of IPR, while some argue for a much greater commons approach with software, Linux being the iconic example (BBC, 2004). It does seem ironic that aggressive protection of advanced nation IPR is accompanied by the rampant exploitation of cultural IPR by multinational organizations (Knapp, 2003; Wired, 2004). Nations such as the U.K. are asset-stripping poorer nations by enticing their expensively trained medics to come and work in the U.K. health service (BBC, 2005a). The approach to IPR protection is uneven, and sometimes hypocritical, and the power relations of IPR and GI will continue to impact significantly on the availability of GI and the tools and techniques to process GI. The early twenty-first century is a time of divergent trends, increasing GI production, some selective censoring of GI, and an increasing monopolization of many essential techniques that people need to process the GI. It promises to be every bit as turbulent as the latter part of the twentieth century.

References

Abascal, J. 2003, March 27–28. Threats and Opportunities of Rising Technologies for Smart Houses. Paper presented at the Conference on "Accessibility for All." Nice, France. http://www.etsi.org/cce/proceedings/3_2.htm (accessed November 6, 2005).

Agbaje, G.I. and J.O.Akinyede. 2005. *NGDI Development in Nigeria: Policy Issues on Information Access and Information Dissemination*. 13. United Nations Economic Commission for Africa, Committee on Development Information, Addis Ababa.

AGI. 2004, April. *A Geographic Information Strategy for England: A Consultation Document*. Association for Geographic Information Working Group, London. http://www.statistics.gov.uk/geography/gag/downloads/GAG_APR2004_05_GI%20strategy.pdf (accessed April 3, 2007).

Amin, A. 2002. Spatialities of globalisation. *Environment and Planning A*, 34: 385–399.

Anon. 2006, March 16. A Revolutionary Vehicle Tracking System: Shadow Tracker™ J2. *Directions Magazine*. http://www.directionsmag.com/press.releases/index.php?duty=Show&id=13827 (accessed March 17, 2006).

AP. 2004, April 13. Winemakers Get Juiced about Tech. AP. http://www.wired.com/news/technology/0,1282,63047,00.html (accessed April 14, 2004).

AP. 2005a, March 27. Amazon Knows Who You Are. AP. http://www.wired.com/news/ebiz/0,1272,67034,00.html (accessed March 27, 2005).

AP. 2005b, April 24. Car Computers Track Traffic. AP. http://www.wired.com/news/autotech/0,2554,67323,00.html (accessed April 25, 2005).

Arvidsson, A. 2004. On the "pre-history of the panoptic sort": mobility in market research. *Surveillance & Society*, 1: 456–474.

Bakyawa, J. 2005, January 6. "Duplicate and Unavailable": How Internet Bypasses Ugandan Farmers. PANOS. http://allafrica.com/stories/200501060817.html (accessed January 8, 2005).

Ball, D. and R. Babbage (Eds.). 1989. *Geographic Information Systems: Defence Applications*. Brassey's Australia, Rushcotters Bay, NSW.

Barry, A. 2001. *Political Machines: Governing in a Technological Society*. Athlone Press, London.

Bassolé, A. 2005. *Integration of Spatial Data Infrastructures into National Information Policies: Linking SDI and NICI Development Processes to Speed up the Emergence of the African Information Society*. 58. UN Economic Commission for Africa, Addis Ababa.

Bauman, Z. 2000. *Liquid Modernity*. Polity Press, Cambridge, UK.

BBC. 2004, October 28. UK Report Says Linux Is 'Viable'. BBC. http://news.bbc.co.uk/1/hi/business/3960025.stm (accessed October 28, 2004).

BBC. 2005a, March 15. NHS 'Taking Away Africa's Medics.' BBC. http://news.bbc.co.uk/1/hi/health/4349545.stm (accessed March 15, 2005).

BBC. 2005b, March 30. Privacy Fears over NHS Database. BBC. http://news.bbc.co.uk/1/hi/health/4392555.stm (accessed MArch 31, 2005).

BBC. 2006a, April 5. Drivers on Edge over Cliff Route. BBC. http://news.bbc.co.uk/1/hi/england/north_yorkshire/4879026.stm (accessed April 21, 2006).

BBC. 2006b, April 7. Sat-Nav Blamed for Village Jams. BBC. http://news.bbc.co.uk/1/hi/england/gloucestershire/4781350.stm (accessed April 21, 2006).

Ben-Atar, D.S. 2004. *Trade Secrets: Intellectual Piracy and the Origins of American Indus-trial Power*. Yale University Press, New Haven, CT.

Berkhout, F. and J. Hertin. 2004. De-materialising and re-materialising: digital tech-nologies and the environment. *Futures*, 36: 903–920.

Bibby, A. 2006, March 12. Left Hanging on the Line as Call Centres Prioritise the Wealthy. AndrewBibby.com. http://www.andrewbibby.com/misc/callcentres. html (accessed March 13, 2006).

Biever, C. 2004, March 17. RFID Chips Watch Grandma Brush Teeth. *New Scientist*. http://www.newscientist.com/news/news.jsp?id=ns99994788 (accessed March 18, 2004).

Blakemore, M. 2004, January. Discourses of Data: Globalisation, Infrastructures, Agendas. *GIM International*, 18. http://www.gitc.nl (accessed January 15, 2004).

Blakemore, M. 2005, September. Surveillance in the Workplace: An Overview of Issues of Privacy, Monitoring, and Ethics. GMB. http://www.gmb.org.uk/Tem-plates/Internal.asp?NodeID=92346 (accessed September 20, 2005).

Blanchette, J.-F. and D.G. Johnson. 2002. Data retention and the panoptic society: the social benefits of forgetfulness. *Information Society*, 18: 33–45.

Boggan, S. 2006, November 17. Cracked It! *Guardian* (London). http://technology. guardian.co.uk/news/story/0,,1950224,00.html (accessed November 19, 2006).

Bolter, J.D. and R. Grusin. 1999. *Remediation: Understanding the New Media*. MIT Press, Cambridge, MA.

Boyd, C. 2004, December 16. Halal Site Guides Hungry Muslims. BBC. http://news. bbc.co.uk/1/hi/technology/4092947.stm (accessed December 18, 2004).

Bulkeley, H. 2001. Governing climate change: the politics of risk society? *Transactions of the Institute of British Geographers, NS*, 26: 430–447.

Clinton, W. 1994. Coordinating Geographic Data Acquisition and Access: The National Spatial Data Infrastructure, Executive Order 12906. *Federal Register*, 59: 17671–17674. http://frwebgate1.access.gpo.gov/cgi-bin/waisgate.cgi?WAISd ocID=5671913174+0+0+0&WAISaction=retrieve (accessed October 18, 2000).

Clout, H. and C. Gosme. 2003. The naval intelligence handbooks: a monument in geographical writing. *Progress in Human Geography*, 27: 153–173.

Coghlan, A. 1994, January 22. Strong ground for Britain's ploughshares. *New Scientist*.

Commons. 2003, September 17. Oral Evidence, Taken before the Treasury Sub-Com-mittee on Wednesday 17 September 2003. House of Commons, Treasury Select Committee. http://www.publications.parliament.uk/pa/cm200203/cmselect/ cmtreasy/uc1112-i/uc111202.htm (accessed October 20, 2003).

Congress. 2004. *The 9/11 Commission Report*. U.S. Congress, Washington, DC.

Craglia, M. and M. Blakemore. 2004. Access models for public sector information: the spatial data context. In *Public Sector Information in the Digital Age. Between Mar-kets, Public Management and Citizens Rights*, G. Aichholzer and H. Burket (Eds.). Edward Elgar Publishing, Cheltenham, U.K., pp. 187–213.

Crampton, J.W. 2003. Cartographic rationality and the politics of geosurveillance and security. *Cartography and Geographic Information Science*, 30: 135–148.

Crang, M. 2000. Public space, urban space and electronic space: would the real city please stand up? *Urban Studies*, 37: 301–317.

Cross, M. 2005, March 17. Land of Opportunity. *Guardian* (London). http://www. guardian.co.uk/online/story/0,,1438826,00.html (accessed May 3, 2005).

Curry, M.R., D.J. Phillips, and P.M. Regan. 2004. Emergency response systems and the creeping legibility of people and places. *Information Society*, 20: 357–369.

Dai, X. 2003. A new mode of governance? Transnationalisation of European regions and cities in the information age. *Telematics and Informatics*, 20: 193–213.

Dangermond, J. 2001, October 9. Grassroots Geography Sets Tone for 21st User Conference: GIS Communities Are Poised to Take the Pulse of the Planet. Environmental Systems Research Institute. http://www.esri.com/news/arc-news/fall01articles/giscommunities.html (accessed February 28, 2003).

DARPA. 2003, May 20. Terrorism Information Awareness. DARPA. http://www.darpa.mil/body/tia/terrorism_info_aware.htm#Question1/ (accessed May 22, 2003).

Defense. 2004. Announcement of intent to initiate the process to remove aeronautical information from public sale and distribution. *Federal Register* 69: 67546.

Dodge, M. and R. Kitchin. 2004. Flying through code/space: the real virtuality of air travel. *Environment and Planning A*, 36: 195–211.

Doyle, P., J.I. Lane, J. Theeuwes, and L. Zayatz (Eds.). 2001. *Confidentiality, Disclosure and Data Access: Theory and Practical Applications for Statistical Agencies.* North Holland/Elsevier, Amsterdam.

Dreaper, J. 2005, December 24. How Technology Is Aiding Medicine. BBC. http://news.bbc.co.uk/1/hi/health/4555314.stm (accessed December 28, 2005).

Earth. 2003, October. Digital Earth. http://www.digitalearth.gov/ (accessed October 30, 2003).

Economist. 2003a, November 20. In the Long Run We Are All Broke. http://www.economist.com/finance/displayStory.cfm?story_id=2227506 (accessed November 20, 2003).

Economist. (2003b). *Something must (not) be done.* (September 11) Economist, [cited September 14, 2003]. http://www.economist.com/opinion/displayStory.cfm?story_id=2052051.

Economist. 2005a, April 28. The No-Computer Virus. http://www.economist.com/business/displayStory.cfm?story_id=3909439 (accessed April 29, 2005).

Economist. 2005b, March 23. Who Rates the Raters? http://www.economist.com/finance/displayStory.cfm?story_id=3786551 (accessed March 24, 2005).

Elliott, V. 2005, January 24. How a Spy in Sky Will Keep an Eu Eye on Those Who Till the Land. *Times* (London). http://www.timesonline.co.uk (accessed January 28, 2005).

ENSI. 2005, February 7. A Tsunami Rethink for Mapping Policy. Express News Service (India). http://cities.expressindia.com/fullstory.php?newsid=116818 (accessed February 10, 2005).

Environment. 1987. *Handling Geographic Information.* HMSO, Department of Environment, London.

Environment. 2003, September. Environment Agency Property Search. Environment Agency. http://www.environment-agency.gov.uk/yourenv/497473/?version=1&lang=_e (accessed September 3, 2003).

Environment Agency. 2003, September 24. Contribution to the Extended Impact Assessment of INSPIRE: Report of the INSPIRE Framework Definition Support Working Group. European Commission. http://inspire.jrc.it/reports/fds_report.pdf (accessed April 10, 2007).

ESRI. 1998. *The Role of Geographic Information Systems on the Electronic Battlefield.* ESRI, Redlands CA.

Eupolitix. 2003, November 13. New EU Borders Agency, Com/2003/687. Eupolitix.com. http://www.eupolitix.com/EN/Legislation/e32e4bf5-8530-4315-87f9-bcd827f38efd.htm (accessed November 20, 2003).

Europe. 2004a. *EU Plan of Action on Combating Terrorism: Update.* 50. European Commission, Brussels.

Europe. 2004b. *Proposal for a Directive of the European Parliament and of the Council Establishing an Infrastructure for Spatial Information in the Community (INSPIRE).* 31. European Commission, Brussels.

Europe. 2005. *Biometrics at the Frontiers: Assessing the Impact on Society.* For the European Parliament Committee on Citizens' Freedoms and Rights, Justice and Home Affairs (Libe). 166. Joint Research Centre (DG JRC), Institute for Prospective Technological Studies, Brussels.

Europe. 2006, November 22. Environmental Data: Commission Welcomes Agreement on Inspire System. European Commission. http://europa.eu.int/rapid/press-ReleasesAction.do?reference=IP/06/1612&format=HTML&aged=0&language=EN&guiLanguage=en (accessed November 23, 2006).

Europe. 2007, April 25. Directive 2007/2/EC of the European Parliament and of the Council of 14 March 2007 Establishing an Infrastructure for Spatial Information in the European Community (INSPIRE). *Official Journal of the European Union,* L 108, 50. http://eur-lex.europa.eu/LexUriServ/LexUriServ.do?uri=OJ:L:2007:108:0001:0014:EN:PDF (accessed April 25, 2007).

Eurostat. 2004. Results Task Force XIA: Extended Impact Assessment of INSPIRE Based on Revised Scope. European Commission, DG Eurostat. http://inspire.jrc.it/reports/inspire_extended_impact_assessment.pdf (accessed April 10, 2007).

Evans, P. and T.S. Wurster. 2000. *Blown to Bits: How the New Economics of Information Transform Strategy.* Harvard Business School Press, Boston.

Experian. 2004, March. The Impact of Credit Cards on Scores. Experian Consumer DirectSM. http://www.nationalscoreindex.com/ScoreNews_Archive_01.aspx (accessed August 24, 2004).

Experian. 2005. Creditexpert Monitoring Service. Experian Consumer DirectSM. https://www.creditexpert.co.uk/UK/MCCLanding.aspx (accessed March 16, 2005).

FGDC. 2004a. *Guidelines for Providing Appropriate Access to Geospatial Data in Response to Security Concerns.* 13. Federal Geographic Data Committee, Washington, DC.

FGDC. 2004b, May 3. Public Review: Security Concerns about Access to Geospatial Data. Federal Geographic Data Committee. http://www.fgdc.gov/whatsnew/whatsnew.html (accessed May 11, 2004).

Fitzpatrick, T. 2002. Critical theory, information society and surveillance technologies. *Information, Communication and Society,* 5: 357–378.

Fornefeld, M. and P. Oefinger. 2001. *Market Survey: Boosting of the Geospatial Data Market in North Rhine Westphalia.* MICUS Management Consulting GmbH, Düsseldorf.

Forrester. 2000, September 11. Global Ecommerce Will Crush Today's Brittle Supply Chains. Forrester Research. http://www.forrester.com/ER/Press/Release/0,1769,394,FF.html (accessed September 12, 2000).

Fox, S., J.Q. Anderson, and L. Rainie. 2005. *The Future of the Internet: In a Survey, Technology Experts and Scholars Evaluate Where the Network Is Headed in the Next Ten Years.* PEW Internet & American Life Project, Washington, DC.

Frank, A. 2002. *The Surveying Activities at the Austrian Federal Office for Metrology and Surveying: An Economic Analysis* [*Vokswirtschaftliche Studie Zu Den Leistungen Des Bundesamtes Für Eich Und Vermessungswesen Im Vermessungswesen*]. 74. Austrian Federal Ministry of Economics and Labor, Vienna.

Fussey, P. 2004. New Labour and new surveillance: theoretical and political ramifications of CCTV implementation in the UK. *Surveillance and Society,* 2: 251–269.

Galperin, H. 2004. Beyond interests, ideas, and technology: an institutional approach to communication and information policy. *Information Society,* 20: 159–168.

GAO. 2004, November. Data Quality: Census Bureau Needs to Accelerate Efforts to Develop and Implement Data Quality Review Standards, GAO-05-86. GAO. http://www.gao.gov/atext/d0586.txt (accessed January 19, 2005).

GAO. 2005. *Homeland Security: Much Is Being Done to Protect Agriculture from a Terrorist Attack, but Important Challenges Remain.* GAO, Washington, DC.

GartnerG2. 2005. *Copyright and Digital Media in a Post-Napster World*. 51. GartnerG2 and the Berkman Center for Internet and Society at Harvard Law School, New York.

GCD. 2003, July. Welcome to the Official GCD Website. Global Cities Dialogue. http://www.globalcitiesdialogue.org/ (accessed July 25, 2003).

GIPanel. 2005, March 16. Work to Date on GI Strategies in the United Kingdom, background paper. Geographic Information Panel (GIP). http://www.gipanel.org.uk/gipanel/docs/GIStrategyPaper.pdf (accessed November 16, 2006).

GITA. 2006. About the Technology. Geospatial Information and Technology Association. http://www.gita.org/about_gita/background.html (accessed November 16, 2006).

Graham, S. and S. Guy. 2003. Digital space meets urban place: sociotechnologies of urban restructuring in downtown San Fransisco. *City*, 6: 369–382.

Graham, S. and D. Wood. 2003. Digitizing surveillance: categorization, space, inequality. *Critical Social Policy*, 23: 227–248.

GSDI. 2003, October. Global Spatial Data Infrastructure. U.S. Geological Survey. http://www.gsdi.org (accessed October 30, 2003).

Guimet, J. 2004. Four rules to set up a basic SDI: the experience of the (local) SDI of Catalonia region. In *Proceedings of the GSDI-7 Conference*, Bangalor, India. http://gsdidocs.org/gsdiconf/GSDI-7/papers/NIjg.pdf (accessed March 22, 2007).

Haggerty, K.D. and R.V. Ericson. 2000. The surveillant assemblage. *British Journal of Sociology*, 51: 605–622.

Hill, K. 2003, February 4. Price Matching Follows Online Customers' Moves. *E-Commerce Times*. http://www.ecommercetimes.com/perl/story/20670.html (accessed February 7, 2003).

Hindu. 2005, March 11. Over 6,000 Villages to Be Linked under E-Gram Project. *Hindustan Times*. http://www.hindustantimes.com/news/181_1274996,0008.htm (accessed March 12, 2005).

Hine, J., D. Swan, J. Scott, D. Binnie, and J. Sharp. 2000. Space: information provision and wayfinding. *Urban Studies*, 37: 1757–1770.

Hirschhausen, C. Von. 1999. What infrastructure policies for post-socialist Eastern Europe? Lessons from the public investment programmes (PIP) in the Baltic countries. *Europe-Asia Studies*, 51: 417–432.

Hollywood, J., D. Snyder, K. McKay, and J. Boon. 2004. *Out of the Ordinary: Finding Hidden Threats by Analyzing Unusual Behavior*. Rand Corporation. Santa Monica, CA.

Home. 2004, September 28. Cutting-Edge Technology to Secure UK Borders for the 21st Century. Home Office. http://www.gnn.gov.uk/content/detail.asp?NewsAreaID=2&ReleaseID=130801 (accessed September 29, 2004).

IPTS. 2003. *Security and Privacy for the Citizen in the Post-September 11 Digital Age: A Prospective Overview*. 188. Institute for Prospective Technological Studies, Brussels.

ISCGM. 2003, October. Global Map. International Steering Committee for Global Mapping. http://www.iscgm.org/html4/index.html (accessed October 30, 2003).

ITV. 2004, October 5. Flood Map Could Slash House Prices. ITV. http://www.itv.com/news/britain_1826934.html (accessed October 5, 2004).

Jayasuriya, K. 2004. The new regulatory state and relational capacity. *Policy and Politics*, 32: 487–501.

Kablenet. 2003, May 13. EU Counts Sheep. Kable News Service. http://www.kablenet.com/kd.nsf/Frontpage/B17ABFF90F350C6D80256D250057FA65?OpenDocument (accessed May 16, 2003).

Kablenet. 2004, July 8. BSE IT Safeguards 'Obsolete.' Kable News Service. http://www.kablenet.com/kd.nsf/Frontpage/51CE42C9B086B80A80256ECB003AE847?OpenDocument (accessed July 9, 2004).

Kablenet. 2005, April 27. Supercomputer Sharpens Forecasts. Kable Government Computing. http://www.kablenet.com/kd.nsf/Frontpage/328732F3ED387F3780256 FF0003BB981?OpenDocument (accessed April 27, 2005).

Kablenet. 2006, February 20. Tag the Patients. Kable Government Computing. http://www.kablenet.com/kd.nsf/Frontpage/0263B10909CFD3688025711B00453 AAF?OpenDocument (accessed February 22, 2006).

Kingston, R. 2005. The unmaking of citizens: banishment and the modern citizenship regime in France. *Citizenship Studies*, 9: 23–40.

Knapp, C. 2003, October 11. UN steps up action on traditional knowledge. *Lancet*, 362. http://www.thelancet.com (accessed February 18, 2004).

Koontz, L.D. 2004. *Geospatial Information: Better Coordination and Oversight Could Help Reduce Duplicative Investments*, statement of Linda D. Koontz, Director, Information Management Issues. Government Accountability Office, Washington, DC.

Kubosova, L. 2005, April 15. Poland to Host EU Borders Agency. *EU Observer*. http://www.euobserver.com/?sid=9&aid=18856 (accessed April 17, 2005).

Landmark. 2003, February 21. Useful Information. Landmark Information plc. http://www.envirosearch.info/furtherinfo.htm#contamination (accessed February 21, 2003).

Lash, S. 2002. *Critique of Information*. Sage, London.

Lavan, M.J. 2003, May 12. Fishkill Appeals Census Revision. *Poughkeepsie Journal*. http://www.poughkeepsiejournal.com/today/localnews/stories/lo051203s4. shtml (accessed May 12, 2003).

Lawrence, V. 2004, January 28–30. The Changing Role of National Mapping Organisations: A Case Study of Ordnance Survey. Paper presented at the MapIndia 2004 Conference. GIS Development. http://www.gisdevelopment.net/policy/international/mi04003.htm (accessed August 23, 2004).

LGA. 2003, July. Implementing Electronic Government Statements. LGA. http://www.localegov.gov.uk/page.cfm?pageID=186&Language=eng (accessed July 25, 2003).

Lievesley, D. 2001. Making a difference: a role for the responsible international statistician? *The Statistician*, 50: 1–38.

Longhorn, R.A. 2000. Regional geographic information policy: fact or fiction: the case in Europe and lessons for GSDI. In *Proceedings of the 4th Global Spatial Data Infrastructure Conference*, Cape Town, South Africa, March 13–15, 2000.

Longhorn, R. and M. Blakemore. 2004. Re-visiting the valuing and pricing of digital geographic information. *Journal of Digital Information*, 4: 1–27.

Mandela, N. 1997. The waning nation-state. In *At Century's End: Great Minds Reflect on Our Time*, N.P. Gardels (Ed.). Wofhound Press, Dublin, pp. 289–295.

Marks, P. 2005, April 3. Electronic Patent Databases Invent Difficulties. *New Scientist*. http://www.newscientist.com/article.ns?id=dn7213 (accessed April 4, 2005).

Marx, G. 2003. Some information age techno-fallacies. *Journal of Contingencies and Crisis Management*, 11: 25–31.

Medford, C. 2006, November 29. Intel, Wal-Mart Plan Health Net. *Red Herring*. http://www.redherring.com/Article.aspx?a=19942&hed=Intel%2c+Wal-Mart+Plan+Health+Net (accessed November 30, 2006).

Merali, Z. 2006, December 8. Invented for the Military, Used to Defend Wildlife. *New Scientist*. http://www.newscientist.com/channel/tech/mg19225816.000-invented-for-the-military-used-to-defend-wildlife.html (accessed December 17, 2006).

Miller, J. 2003, June 10. OMB: Rein-In Spending on Geospatial Systems. *Government Computer News.* http://www.gcn.com/vol1_no1/daily-updates/22372-1.html (accessed June 13, 2003).

MOD. 2006. Mission: Deliver Geoint, Geospatial Information, Services and Liaison. MOD. http://www.mod.uk/DefenceInternet/AboutDefence/WhatWeDo/Securityand Intelligence/DIS/ICG/DefenceGeographicCentre.htm (accessed November 16, 2006).

MORI. 2004. *E-Citizenship: What People Want,* research study conducted by MORI for the E-Citizen National Project. MORI, London.

Mosco, V. 2004. *The Digital Sublime: Myth, Power and Cyberspace.* MIT Press, Cambridge, MA.

Multimap. 2001, October 1. Multimap.com Awarded US Internet Patent. Multimap.com. http://www.multimap.com/indexes/pressindex.htm (accessed March 23, 2002).

Nag, P. 2002, February. Maponomics. Survey of India. http://www.gisdevelopment. net/gismarket/gismarket002.htm (accessed March 23, 2002).

Naughton, J. 1999. *A Brief History of the Future: The Origins of the Internet.* Orion Books, London.

NGPO. 2005, April. National Geospatial Programs Office. U.S. Geological Survey. http://www.usgs.gov/ngpo/index.html (accessed April 9, 2005).

NIC. 2004, December. Mapping the Global Future. National Intelligence Council (USA). http://www.cia.gov/nic/NIC_globaltrend2020.html (accessed February 10, 2005).

Norwich. 2004, March 3. Norwich Union's Revolutionary Flood Map Begins Roll-Out. Norwich Union plc. http://www.aviva.com/news/release.cfm?section=m edia&filter=none&ID=1684 (accessed March 3, 2004).

ODPM. 2003. *Implementing Electronic Government Return 2003 (IEG3): Proforma.* 17. ODPM, London.

OECD. 1996. Global Information Infrastructure: Global Information Society (GII-GIS). Statement of Policy Recommendations Made by the ICCP Committee, OCDE/GD(96)93. Organization for Economic Co-operation and Development. http://www.oecd.org/dataoecd/3/58/1896739.pdf (accessed July 1, 2004).

OMB. 1990, October 19. Coordination of Surveying, Mapping, and Related Spatial Data Activities. OMB. http://www.whitehouse.gov/omb/circulars/a016/a016. html (accessed June 14, 2002).

OMB. 2006. Geospatial One-Stop. OMB. http://www.whitehouse.gov/omb/egov/c-2-1-geo.html (accessed November 16, 2006).

Onsrud, H.J. 1998. The tragedy of the information commons. In *Policy Issues in Modern Cartography,* D.R.F. Taylor (Ed.). Pergamon, Oxford, pp. 141–158.

PEW. 2005a, January 9. Technology Experts and Scholars Foresee a Bigger Role for the Internet in People's Personal and Work Lives in the Next Decade. PEW Internet & American Life Project. http://www.pewinternet.org/press_release.asp?r=95 (accessed April 9, 2005).

PEW. 2005b, January 25. Trends 2005: A Look at Changes in American Life. PEW Internet & American Life Project. http://www.pewinternet.org/PPF/r/97/ press_release.asp (accessed February 4, 2005).

Pickles, J. (Ed.). 1995. *Ground Truth: The Social Implications of Geographic Information Systems.* Guilford Press, New York.

Pickles, J. 2003. *A History of Spaces: Cartographic Reason, Mapping and the Geo-Coded World.* Routledge, London.

PIRA. 2000. *Commercial Exploitation of Europe's Public Sector Information.* 132. PIRA International, Leatherhead, Surrey, U.K.

Purdam, K., E. Mackey, and M. Elliott. 2004. The regulation of the personal: individual data use and identity in the UK. *Policy Studies*, 25: 267–281.

Reichardt, M. and J. Moeller. 2000. SDI Challenges for a New Millennium: NSDI at a Crossroads: Lessons Learned and Next Steps. March Federal Geographic Data Committee. http://130.11.63.121/docs2000/capetown/GSDIPaperSAv4_v6_jm.doc (accessed May 5, 2005).

Reuters. 2002, September 5. Internet Freedom Also Victim of Sept 11, Group Says. Reuters. http://www.reuters.com/news_article.jhtml?type=internetnews&StoryID=1415602 (accessed September 6, 2002).

Rhind, D. 1996. Economic, legal and public policy issues influencing the creation, accessibility and use of GIS databases. *Transactions in GIS*, 1: 3–12.

Rhind, D., B.W.D. Briggs, and J. Wiggins. 1976. The creation of an environmental information system for the European Community. *Nachrichten aus dem Karten- und Vermessungswesen*, II: 147–157.

Richardson, D.E. 1977. The cruise missile. *Electronics and Power*, 23: 896–901.

Rivers, T.J. 2002. Progress and technology: their interdependency. *Technology in Society*, 24: 503–522.

Roberts, A. 2004. Orcon creep: information sharing and the threat to government accountability. *Government Information Quarterly*, 21: 249–267.

Robins, K. and F. Webster. 1999. *Times of the Technoculture: From the Information Society to the Virtual Life*. Routledge, London.

Rose, J. 2001. *The Intellectual Life of the British Working Classes*. Yale University Press, New Haven, CT.

Sawhney, H. 2001. Dynamics of infrastructure development: the role of metaphors, political will and sunk investment. *Media, Culture & Society*, 23: 33–51.

Sawhney, H. 2003. Wi-fi networks and the rerun of the cycle. *Info*, 5: 25–33.

Schaefer, M. and J. Moeller. 2000, March. Support for International Infrastructure Activities. Federal Geographic Data Committee. http://www.fgdc.gov/international/sha1.pdf (accessed May 5, 2005).

Schmid, G., D. Haynes, and P. Clegg. 2003. Information management as an enabler: a case study from Medway Council. In *Proceedings of the AGI Conference GeoSolutions 2003*, September 16–18, 2003. Association for Geographic Information, London.

Schmidt, H. 2003, November 21. Comparing Prices on the Internet Can Bring Savings of 30 Percent. Frankfurter Allgemeine Zeitung. http://www.faz.com/IN/INtemplates/eFAZ/docmain.asp?rub={F040FFD3-897B-46DF-9603-752DD6405389}&doc={722DCB6C-4B8A-4D37-B1C2-94EDB586FD30} (accessed March 4, 2004).

Schuurman, N. 2000. Trouble in the heartland: GIS and its critics in the 1990s. *Progress in Human Geography*, 24: 569–590.

Scotland. 2006, January 24. One Scotland: One Geography. Scottish Executive. http://www.scotland.gov.uk/Topics/Government/Open-scotland/OneScotland/Introduction/Q/Zoom/80 (accessed November 16, 2006).

Sennett, R. 2006. *The Culture of the New Capitalism*. Yale University Press, New Haven, CT.

Sewell, G. and J.R. Barker. 2001. Neither good, nor bad, but dangerous: surveillance as an ethical paradox. *Ethics and Information Technology*, 3: 183–196.

Shenk, D. 1997. *Data Smog*. HarperCollins, San Francisco.

Singer, E. 2006, November 20. Driving a Wheelchair with Your Shirt. Technology Review. http://www.techreview.com/BioTech/17803/ (accessed November 21, 2006).

Slevin, J. 2000. *The Internet and Society*. Polity Press, London.

Smith, C. and K. Stewart. 2003, October 1. Census Blooper Costly to Utahns. *Salt Lake Tribune*. http://www.sltrib.com/2003/Oct/10012003/utah/97469.asp (accessed October 7, 2003).

Standage, T. 1998. *The Victorian Internet*. Wiedenfeld and Nicolson, London.

Statistics. 2001. *Note of a Meeting on Census Disclosure Control: 13 Dec 2001 at Ons Drummond Gate*. 3. Statistics Commission, London.

Statistics. 2003a. *The 2001 Census in Westminster: Interim Report*. Statistics Commission, London.

Statistics. 2003b, November 4. Census Matching Project for Manchester. National Statistics. http://www.statistics.gov.uk/cci/nugget.asp?id=590 (accessed November 4, 2003).

Stiglitz, J.E. 2002, January 1. Globalism's Discontents. *American Prospect*. http://www.prospect.org/print/V13/1/stiglitz-j.html (accessed January 28, 2003).

Stip, E. 2005. Environmental cognitive remediation in schizophrenia: ethical implications of "smart home" technology. *Canadian Journal of Psychiatry*, 50. http://www.cpa-apc.org/Publications/Archives/CJP/2005/April/ORStip.asp (accessed November 6, 2005).

Survey. 2001, March 5. Joint Statement by Centrica and Ordnance Survey. Ordnance Survey. http://www.ordsvy.gov.uk/ (accessed March 6, 2001).

Survey. 2004. Intellectual Property Policy Statement. Ordnance Survey. http://www.ordnancesurvey.co.uk/oswebsite/business/copyright/policy.html (accessed February 19, 2004).

Survey. 2005, April 6. Performance Benchmarks Set for Ordnance Survey. Ordnance Survey. http://www.ordnancesurvey.co.uk/oswebsite/media/news/2005/April/performancetargets.html (accessed April 8, 2005).

Tapscott, D. 2004, March. The Transparent Burger. Wired.com. http://www.wired.com/wired/archive/12.03/start.html?pg=2?tw=wn_tophead_7 (accessed March 5, 2004).

Telecities. 2003. October. Telecities: Cities Connect. Telecities. http://www.telecities.org/ (accessed October 30, 2003).

Thrift, N. 1997. The rise of soft capitalism. In *An Unruly World? Globalisation and Space*, A. Herod, S. Roberts, and G. Toal (Eds.). Routledge, London, pp. 25–71.

Triggle, N. 2006, May 19. How Gadgets Will Aid Elderly at Home. BBC. http://news.bbc.co.uk/2/hi/health/4996890.stm (accessed May 22, 2006).

Tuomi, I. 2002, November. The Lives and Death of Moore's Law. *First Monday*, 7. http://firstmonday.org/issues/issue7_11/tuomi/index.html (accessed November 15, 2002).

Tuomi, I. 2004, July. Economic Productivity in the Knowledge Society: A Critical Review of Productivity Theory and the Impacts of ICT. *First Monday*, 9. http://firstmonday.org/issues/issue9_7/tuomi/index.html (accessed July 23, 2004).

Twist, J. 2004, August 30. Peering beyond the Technology Hype. BBC. http://news.bbc.co.uk/1/hi/technology/3577746.stm (accessed August 31, 2004).

Umashankar, C. 2005, January. E-Governance: The Key Players. http://www.egovonline.net/jan_feb_05/notebook3.htm (accessed March 14, 2005).

UNECA. 2005a, April. Committee on Development Information.UN Economic Commission for Africa. http://www.uneca.org/codi (accessed April 29, 2005).

UNECA. 2005b. Concept Paper: Codi IV. Information as an Economic Resource. 8. United Nations Economic Commission for Africa, Committee on Development Information, Addis Ababa.

UNECE. 1992. *The Fundamental Principles of Official Statistics in the Region of the Economic Commission for Europe*. 2. Economic Commission for Europe, Geneva.

UNECE. 2001. *Statistical Data Confidentiality in the Transition Countries: 2000/2001 Winter Survey.* Work Session on Statistical Data Confidentiality (Skopje), UNECE Statistical Division. http://www.unece.org/stats/documents/2001.3.confidentiality. htm (accessed May 19, 2001).

UNGIWG. 2007, February. *Strategy for Developing and Implementing a United Nations Spatial Data Infrastructure in Support of Humanitarian Response, Economic Development, Environmental Protection.* UNGIWG, Geneva.

Urry, J. 2003. *Global Complexity.* Polity Press, Cambridge, U.K.

USPTO. 2001a, May 29. Computer System for Indentifying Local Resources (Definitions), U.S. Patent 6,240,360. USPTO. http://patft.uspto.gov/netacgi/nph-Parse r?Sect1=PTO1&Sect2=HITOFF&d=PALL&p=1&u=/netahtml/srchnum.htm&r =1&f=G&l=50&s1=%276240360%27.WKU.&OS=PN/6240360&RS=PN/6240360 (accessed March 23, 2002).

USPTO. 2001b, May 29. Computer System for Indentifying Local Resources (Description), U.S. Patent 6,240,360.USPTO. http://patft.uspto.gov/netacgi/nph-Parser ?Sect1=PTO1&Sect2=HITOFF&d=PALL&p=1&u=/netahtml/srchnum.htm&r =1&f=G&l=50&s1=%276240360%27.WKU.&OS=PN/6240360&RS=PN/6240360 (accessed March 23, 2002).

USPTO. 2005a, March 8. Programmatically Calculating Paths from a Spatially-Enabled Database, U.S. Patent 6,865,479. USPTO. http://patft.uspto.gov/ netacgi/nph-Parser?Sect1=PTO2&Sect2=HITOFF&p=1&u=/netahtml/search-bool.html&r=8&f=G&l=50&col=AND&d=ptxt&s1='geographic+information' &OS=%22geographic+information%22&RS=%22geographic+information%22 (accessed April 4, 2005).

USPTO. 2005b, March 8. Meeting Location Determination Using Spatio-Semantic Modeling, U.S. Patent 6,865,538. USPTO. http://patft.uspto.gov/netacgi/nph-Parser?Sect1=PTO2&Sect2=HITOFF&p=1&u=/netahtml/search-bool.html&r= 42&f=G&l=50&col=AND&d=ptxt&s1=longitude&OS=longitude&RS=longitu de (accessed April 4, 2005).

USPTO. 2005c, March 15. System and Method for Geographical Indexing of Images, U.S. Patent 6,868,169. USPTO. http://patft.uspto.gov/netacgi/nph-Parser?Sect1 =PTO2&Sect2=HITOFF&p=1&u=/netahtml/search-bool.html&r=34&f=G&l= 50&col=AND&d=ptxt&s1=longitude&OS=longitude&RS=longitude (accessed April 4, 2005).

USPTO. 2005d, March 15. Map Data Providing Apparatus, Map Data Installing Terminal Device, and Communication-Type Navigation Apparatus, U.S. Patent 6,868,334. USPTO. http://patft.uspto.gov/netacgi/nph-Parser?Sect1=PTO2&Sec t2=HITOFF&p=1&u=/netahtml/search-bool.html&r=28&f=G&l=50&col=AND &d=ptxt&s1=longitude&OS=longitude&RS=longitude (accessed April 4, 2005).

USPTO. 2005e, March 15. Method of Converting Geospatial Database into Compressive Database for Multiple Dimensional Data Storage, U.S. Patent 6,868,421. USPTO. http://patft.uspto.gov/netacgi/nph-Parser?Sect1=PTO2&Sect2=HITO FF&p=1&u=/netahtml/search-bool.html&r=19&f=G&l=50&col=AND&d=ptxt &s1=longitude&OS=longitude&RS=longitude (accessed April 4, 2005).

USPTO. 2005f, March 22. Intelligent Road and Rail Information Systems and Methods, U.S. Patent 6,871,137. USPTO. http://patft.uspto.gov/netacgi/nph-Parser ?Sect1=PTO2&Sect2=HITOFF&p=1&u=/netahtml/search-bool.html&r=12& f=G&l=50&col=AND&d=ptxt&s1=longitude&OS=longitude&RS=longitude (accessed April 4, 2005).

USPTO. 2005g, March 29. Method of Inputting a Destination into a Navigation Device, and Navigation Database, U.S. Patent 6,873,906. USPTO. http://patft.uspto.gov/netacgi/nph-Parser?Sect1=PTO2&Sect2=HITOFF&p=1&u=/netahtml/search-bool.html&r=2&f=G&l=50&col=AND&d=ptxt&s1=longitude&OS=longitude&RS=longitude. (accessed April 4, 2005).

USPTO. 2005h, March 29. Method and System for Controlling Presentation of Information to a User Based on the User's Conditio, U.S. Patent 6,874,127. USPTO. http://patft.uspto.gov/netacgi/nph-Parser?Sect1=PTO2&Sect2=HITOFF&p=1&u=/netahtml/search-bool.html&r=1&f=G&l=50&col=AND&d=ptxt&s1='geographic+information'&OS=%22geographic+information%22&RS=%22geographic+information%22 (accessed April 4, 2005).

USSS. 2006, November. National Threat Assessment Center: Insider Threat Study. USSS. http://www.secretservice.gov/ntac_its.shtml (accessed November 22, 2006).

Wakefield, A. 2004. The public surveillance functions of private security. *Surveillance & Society*, 2: 529–545.

Weigelt, M. 2006, November 28. Taxpayers Shy from Sharing Info Online. *Federal Computer Week*. http://www.fcw.com/article96930-11-28-06-Web (accessed December 1, 2006).

Wikipedia. 2006, November 15. Anthony Giddens. http://en.wikipedia.org/wiki/Anthony_Giddens (accessed November 27, 2006).

Wired. 2004, February 12. Keep Your Patents Off Our Plants. Wired.com. http://www.wired.com/news/technology/0,1282,62266,00.html (accessed February 12, 2004).

chapter six

Spatial data infrastructures:
Policy, value, and cost–benefit

6.1   Introduction to policy in spatial data infrastructure

Among the key policy issues affecting geographic information (GI) globally are information ownership, custodianship, and preservation; access and exploitation rights; and charging regimes for public sector information (PSI). Some of these issues were examined in earlier chapters. In this chapter, we explore the role of geographic information policies and their implementation strategies within spatial data infrastructure (SDI) and under the umbrella framework of national information infrastructure (NII). In doing so, we revisit the concepts of value of GI and how the many values identified in Chapter 2 affect infrastructure-wide impact assessments or cost–benefit analyses for SDI implementations.

Following the practice of earlier chapters, we begin at the elementary level of defining some basic terms, such as policy, information policy, and strategy, and then present a sample of SDI definitions to see where policy falls within these definitions. This chapter is not meant to be a compendium of SDIs that are evolving around the globe, which has been the focus of several publications over the past decade (Burrough and Masser, 1998; Groot and McLaughlin, 2000; Van Loenen and Kok, 2004; Masser, 2005, 2007; Van Loenen, 2006; Crompvoets, 2006; Onsrud, 2007). Rather, we present samples of SDI initiatives at the national and regional level to provide insight into how policy issues are at the heart of SDI visions, goals, and strategies, along with other technical and organization issues where policies may have only an indirect impact. Many SDI policies are aligned to national information infrastructure (NII) policies, inherently or on purpose, since much GI is in the public sector, and is the the focus of many NII initiatives, including PSI reuse and e-governance.

We start by asking what policies are and why have them. According to the *American Heritage Dictionary,* a policy is a plan of action "intended to influence and determine decisions, actions, and other matters" or a "guiding principle, or procedure considered expedient, prudent, or advantageous." Wikipedia refers to policy as both a thing and a process that "includes the identification of different alternatives, such as programs or spending priorities, and choosing among them on the basis of the impact they will have." Interestingly, infrastructures and especially SDIs have also been labeled both as things (products that exist or are created) and as processes (by which the things are created).

One way of looking at SDI policy might be to see what type of policy it constitutes, for example, distributive, redistributive, regulatory, or constituent-based. Understanding what type of policy is being determined may help also to understand the functional goals of the policy from the viewpoint of the policy makers. Distributive policies extend goods and services to members of an organization or society, as well as distributing the costs of the goods and services among the members of that organization or society. Redistributive policies have the positive impact of distributive policies while simultaneously taking away benefits from other stakeholders. Regulatory policies place limits on organizations or individuals by allowing or disallowing certain behaviors, or otherwise enforcing certain types of good behavior. Examples in the information sector include regulations dealing with intellectual property protection or personal privacy protection. For a regulatory policy to be effective, it must be possible to identify the good behavior and regulate or enforce sanctions for bad behavior. Unfortunately for the SDI policy maker, the types of policies embodied in an SDI strategy could place the SDI policy in almost any one of these types, and sometimes in more than one type simultaneously.

Burger (1993, p. 18) states that constituency-based policies are the most difficult to characterize or describe, quoting Salisbury (1968, p. 158) who contends that they impose constraints on a group but are perceived to increase and not decrease benefits to the group. Lowi's (1972) definition of constituent policy confers broad costs and benefits to society assuming a top-down process of policy making dominated by elected officials and administrative agencies, as opposed to policy that affects narrow, often economic, interests. Tolbert (2002) refined this concept to include governance policy, which "has a prominent procedural component and can be initiated by a bottom-up process of policymaking, via citizen initiatives or interest groups, as well as by a top-down process through political elites."

Wikipedia proposes that constituent policies create executive powers or deal with laws. For example, in the Spanish province of Catalonia, Law 16/2005 of December 2005 creates executive powers for a regional cartographic commission and places responsibilities on the regional cartographic institute regarding GI and SDI for the province. This is an example of a constituent policy setting out goals and responsibilities. A separate decree in October 2006 sets the regulations by which the policy in the law is to be enacted and enforced, which is an example of regulator policy that includes concrete action plans.

We look at policy as a product in section 6.1.2 and as a process in section 6.1.3. First, let us look more closely at information policy itself, since the main policy element in any SDI relates to the information. We will not investigate further the distinctions between information policy and knowledge policy proposed by Bawden (1996), except to note his conclusion that information policy is "dependent upon an appreciation of the meaning and significance of knowledge in its context."

6.1.1   Information policy

What is information policy, and what is unique about it compared to other types of policy? According to Burger (1993), information policy is but one of many types of public policy, yet is seldom mentioned specifically or separately in public policy literature reviews prior to 1980. In the 1990s, information policy was usually lumped in with information and communications technology (ICT) policy, including information management. While many of the main issues in ICT policy are relevant, information policy also includes "much more, such as scientific and technical information policy, privacy issues, literacy, freedom of speech, libraries and archives, secrecy and its effects on commercial information policy and national security, and access to government information" (Burger, 1993, p. 3). Burger proposes three reasons for apparent difficulty in understanding information policies, the first of which is that "information remains an intangible enigma" (Burger, 1993, p. 5) despite the considerable research and resources expended on such understanding, multiple definitions, often unquantifiable benefits, etc. His second reason is that information policy deals with policy, which he acknowledges is not a particularly remarkable insight, but notes that even political scientists who deal extensively in policy issues have difficulty defining and understanding policy, so why should information policy be any different. His final reason is that information is pervasive, "involved in every social choice we make" — how similar to the oft-quoted "GI is everywhere" proclamation of the GI community.

Rowlands (1996, p. 11) notes that information policy is characterized by:

- Involvement of large numbers of stakeholders (a result of the ubiquity of information).
- Information policy decisions may impact on other events and policies in numerous other sectors than that for which the policy was first defined.
- It is difficult to use traditional policy analysis methodologies where information is concerned.
- Information policy is made at many different levels, from private and organizational up through all levels of government, even globally.

Different information policies also depend upon the type of information that is the focus of the policy, e.g., private vs. public, and how the information is to be used, i.e., as a public good or a tradable commodity, available via unrestricted information flow vs. closed, restricted flow, e.g., via strong intellectual property rights (IPR) protection or other (Rowlands 1996, p. 15). This level of complexity gives rise to naturally occurring contests between how different types of information is disseminated and used, as discussed in Chapters 3 to 5.

Regarding information policy goals, we will see that SDI policy goals are not that different from those of other major government information policies. For example, the U.S. National Commission on Libraries and Information Science (NCLIS), established by law in 1970, is a permanent, independent agency of the federal government that advises the president and Congress on the implementation of policy affecting libraries and information provision generally. In response to the threatened closure of the National Technical Information Service (NTIS) in the Department of Commerce in 1999, at the request of U.S. congressional leaders, NCLIS launched a study into "fundamental issues regarding how the government used, disseminated and valued its information resources" (NCLIS, 2001, p. 3). The report was produced and widely circulated within federal agencies, including by the Office of Management and Budget (OMB). The Commission proposed 36 recommendations, 16 of which were classed as strategic. These fell into the following main categories:

- Creating three new federal government-level offices responsible for different types of information plus retaining the NTIS (and its budget)
- Implementing a separate information dissemination budget
- Strengthening existing federal acts and regulations relating to information dissemination by and within federal agencies
- Encouraging similar moves at state and local government levels
- Fostering stronger partnering with the private sector, especially for value-added products and services
- Better coordination at the federal government level
- Greater training and awareness activities plus improved access technology for greater inclusion of civil society

In the recommendations listed above, the reader familiar with SDI strategies can see direct parallels with similar policy goals and recommendations at the national and regional level regarding SDI creation, which will become more apparent in section 6.2.

6.1.2   Policy as product

Formal policy statements are the means by which policy makers define specific goals for their policies, which can be political, financial, administrative, or operational. Goals can also be classified as economic, societal, socioeconomic, or governance related. Policy as a product is often embodied in model policies that are promulgated by either law or regulation, or as some other form of official recommendation, the latter typically not as enforceable as the former. Model policies or policy statements usually comprise a justification for needing the (new) policy, the rationale behind the policy proposed in the model or statement, and references to goals and (perhaps) success criteria (if evaluation of the policy is mentioned in the document). Policy statements or model policies need not specify actual implementation procedures or actions, since many different approaches may be employed to achieve the policy's goals, and these implementation measures and associated instruments may change over the timescale that the main policy remains in effect.

Orna (1999) proposed a range of components for an organization's information policy, which we feel apply equally to the information policy elements within a national or regional SDI, including:

- Stating the overall objectives for information use in the organization and priorities within these objectives
- Defining what constitutes information in regard to the policy
- Defining information management principles
- Defining human resource management principles
- Proposing technology to use to support information management for achieving the policy goals
- Defining cost-effectiveness principles for both information and knowledge management

Those readers familiar with the European Union's INSPIRE directive (EU, 2007) will note the striking similarity between the information policy components listed above and those found in the principle articles of the directive relating to a pan-European SDI.

SDI policies relate primarily to government information issues and are thus a subset or special application of wider public policy planning, of public sector information (PSI) policy, and e-government policies and strategies. This overlap is due to the oft-quoted maxim that "GI is everywhere." Since public sector GI (PSGI) is both public sector information and geographic information, it is virtually impossible that SDI can be defined and created without intersecting with NII policies and strategies.

It is often difficult to separate the policy product from the policy process. For example, research in Scotland into model policies for land use planning started with the premise that the study was "as much concerned with the processes involved in preparing and maintaining model policies as the policies themselves. It thus deals with policy as product and policy as process" (Scottish Executive, 2004). The Scottish Executive found that model policies that focused on words, form, style, and content in order to compare different land development practices suffered from too great an emphasis on the product — the model policy wording — which "may not be sufficiently sensitive to the wider policy processes required to sustain model policies" (Scottish Executive, 2004, p. 19).

6.1.3   Policy as process

Rajabifard (2002) recommended "adoption of an SDI process-based model instead of the current strategy for the APSDI development ... a better approach to overcome some of the challenges facing SDI initiatives persisting with a product-based approach, especially in this region," based on the innovation process model of Rogers (1995), since innovation and infrastructure creation have many similarities. Viewing SDI policy as a process vs. a product is useful because of the complex interactions among social, economic, and political issues that are inherent to SDI formulation. Policies are made and implemented in the same way that decisions are made and implemented. However, not all actions that implement policies are necessarily considered to be a part of the policy itself, since a policy can be implemented in many different ways, and allied actions may result that are beneficial but not policy oriented.

Copeland and Antikarov (2001) present another view on decision making as a process by which different real options can be explored, and relevant options selected and then implemented. Yet their practitioner's guide does not delve into policy issues, since the real options methodology described can be used in relation to implementing any number of different policies. Thus, the distinction between policy as product and policy as process should be noted throughout the different phases of policy definition vs. implementation. Also, note an important distinction between policy makers and decision makers, in that there are relatively few of the former compared to the latter, and decision makers must operate within the policies set by the policy makers.

For practitioners of policy making, a policy may be like a decision, but "it is not just a 'one-off', independent decision"; rather, it is a "set of coherent decisions with a common long-term purpose" (ILRI, 1995). Policies progress from agreed statements of goals and principles to the actions implemented to achieve those goals, following strategies, plans, programs, and finally specific projects or enforcement of (new or modified) laws or regulations, whether in government or within an organization. In many jurisdictions, both national and regional (or transnational, such as the EU), it is common for a law (or directive in the case of the EU) to be the mechanism that expresses and legitimizes a policy, while a separate set of regulations or decree or similar mechanism (recommendation or council decision in the EU) specifies the processes by which the law is to be enacted.

The policy process has received various treatments by different authors and practitioners. Burger (1993, pp. 8–17), drawing on Kelman (1987), proposed three main stages:

- Policy formation, which produces the policy goals and instructions, including the initial proposal, based on some rationale for action and evaluation of that rationale and proposed goals
- Policy implementation, which includes legislation and implementing measures
- Postimplementation evaluation, which Burger claims is not always as rigorous as might be desired since some policy makers have personal stakes in the policies they promoted, and thus may not be keen

to have these policies scrutinized too closely later on, in regard to goals achieved or resources consumed

Expanding upon Bridgman and Davis (2004), a more useful policy process cycle (PPC) model that better reflects what we see in practice in information policy development might be something like:

1. Identify issues that are the focus of the policy being developed, including rationale (why action is needed) and expected goals or results (positive impact on the organization or society).
2. Identify proposed policy instruments to enable implementation, taking into account instruments that may already exist, e.g., prior information legislation regarding intellectual property rights (IPR), licensing regimes, data access or reuse regulations, etc.
3. Analyze alternatives to the policy instruments and examine the potential impact of the alternative instruments on achieving the policy's goals, at what cost, to whose benefit, etc.
4. Identify and consult with major stakeholders on the draft policy and instruments, including alternatives. Stakeholder involvement is crucial here and in the following steps.
5. Make the final decisions among alternatives, e.g., regarding principles, implementation instruments, enforcement procedures and practices.
6. Implement the policy via the agreed-upon mechanisms, taking into account existing legal instruments relating to information policy.
7. Perform postimplementation evaluation of the impact of the policy. Based on the evaluation, revisit the cycle from step 1.

It is worth noting that while most SDI initiatives have progressed at least to step 4 in regard to policy implementation, many are still trapped in step 5, and only few will claim to have completed step 6 (full implementation). Thus, none have yet reached step 7 — evaluation and subsequent reinvestigation of the original goals, policies, and instruments. One exception might be the U.S. National Spatial Data Infrastructure (NSDI) (described later), within which the "product" (the national SDI) — more than a decade since its definition and authorization for implementation by executive order in April 1994 — was seen by many as not achieving its original stated goals (Corle, 2004; Koontz, 2004; Longhorn, 2006) due mainly to lack of sufficient participation in the national initiative by academia, local and state government, and private industry. In a review of impact analyses or cost–benefit studies for SDIs globally, none have yet been found (by the authors) which relate to evaluating an existing SDI. Appropriate indicators of the potential success (or failure) of an SDI implementation are the focus of ongoing research that we do not expect to be completed for some years (Crompvoets, 2007).

The PPC policy-making process presumes a coordination activity that begins with step 1 and runs through step 7. Coordination implies an owner for the initiative or policy definition and implementation process. For SDI formation, the owner is not always obvious, once again because of the claimed ubiquity of GI, especially for multiple government agencies who both create and use GI, and the overlap of GI with other, higher-level information policies, such as NII, e-government, and e-commerce. Lack of an appropriate owner or change of owner midstream of the policy definition and implementation process can be fatal or, at the very least, can delay the whole process for a number of years. We saw this in Europe in 1999 regarding the GI2000 initiative, which began as an "information market" action in DG Information Society — the first attempt at a pan-European SDI — and which was cancelled after 5 years of effort. This was followed 8 years later by the successful adoption of the INSPIRE directive creating such an SDI, under joint sponsorship and ownership of the DG Joint Research Center, DG Environment, and DG Eurostat. These three DGs all had a greater need for joined-up GI across Europe for regional planning, monitoring, and enhanced governance activities than did the DG Information Society, where GI played a relatively small part in the existing European multimedia information marketplace.

6.2   Examples of SDI developments at national and regional levels

Although spatial data infrastructure (SDI) was discussed in Chapter 5 in relation to wider public sector information (PSI) issues, including governance, SDI has not yet been defined. In fact, there are a number of different definitions for SDI extant, although they all have many similar characteristics depending on the national and institutional context. Some of the different definitions for SDI are presented here, at national, regional, and global levels, and from both historical and current viewpoints.

6.2.1   SDI developments in the U.K.

Discussions concerning an SDI for the U.K. began mid-1995 following a lead from the European Commission earlier that year with its GI2000 initiative for a pan-European SDI that would be based on interconnected national-level SDIs, now embodied in the Infrastructure for Spatial Information in Europe (INSPIRE) directive (EU, 2007). The first-pass U.K. SDI proposed creating a U.K. National Geospatial Data Framework (NGDF). This framework would facilitate unlocking national GI resources by enabling greater awareness of data availability, improving access to the data, and integrating data through use of standards. NDGF was not intended to create a physical framework or to deliver data sets, services, or products, but its use was expected to facilitate value-added services by enabling the combination of data from multiple sources, from both the private and public sectors (NGDF Management Board, 1999).

Then, in 2000, the emphasis shifted to the Digital National Framework (DNF), defined as:

> a model for the integration of geographic information of all kinds ... supported by a set of enabling principles and operational rules that underpin and facilitate the integration of geo-referenced information. (Ordnance Survey, 2004, p. 13)

The main principles embodied in the DNF include:

- DNF concepts and methods should meet the strategic needs of the whole GI community.
- Data should be collected only once and then reused.
- Reference data (core GI) should be captured at the highest resolution practical, so that it can be more widely reused to "meet analysis and multi-resolution publishing requirements." (Ordnance Survey, 2004, p. 13)
- Existing *de facto* and *de jure* standards will be used wherever possible.

Key DNF goals to help realize the benefits of applying the DNF model include:

- Establishing a coherent structural model of national reference data sets and relationships with application information
- Creating and maintaining a national information framework based on this model to support consistent integration of GI and enable true interoperability
- Evolving a consistent approach to georeferencing and establishing consistent interrelationships between reference data and application data

As the U.K.'s Digital National Framework continues to evolve, the scope is expected to expand to include a model of the relationships among key national GI data sets, technical support to enable GI interoperability, and greater dialogue and cross-sectoral communication. One example is the work within the hydrographic community to extend the DNF to include offshore GI, being promoted and enabled by the U.K. Hydrographic Office and its commercial subsidiary SeaZone Solutions Ltd. (Osborne and Pepper, 2006). Three regional (subnational) SDIs have been created in the U.K. — in Wales (AGI Cymru, 2003), Scotland (Scottish Executive, 2005), and Northern Ireland (OSNI, 2002) — yet, as of June 2007, there was no national GI or SDI strategy other than the DNF, which is only one component of an SDI and is not presented as a complete SDI. A GI strategy for the U.K. is being considered, following a study (unpublished publicly) completed for the GI panel, a U.K. government advisory body, in December 2006 (GI Panel, 2007).

The U.K. does have a reasonably well-developed e-government information infrastructure, with established standards for both an e-government interoperability framework and an e-government metadata system. National legislation exists that implements the EU's directive on Re-Use of Public Sector Information, as does a Freedom of Information Act. Databases are protected by the EU directive on legal protection of databases adopted across all EU member states in 1996.

6.2.2   SDI developments in the U.S.

In the U.S., the NSDI concept first launched in April 1994 by executive order (Clinton, 1994) has evolved into a wider framework approach as "a means to assemble geographic data nationwide to serve a variety of users ... a collaborative community based effort in which these commonly needed data themes are developed, maintained, and integrated by public and private organizations within a geographic area" (FGDC, 2007a). The framework:

- Forms the GI backbone of the NSDI, with the overall objective of permitting local, regional, state, and federal government organizations and private companies to share resources, improve communications, and increase efficiency
- Comprises the most commonly needed and used GI, procedures, and technology for building and using the data, and institutional relationships and business practices that support the environment
- Is expected to facilitate production and wider use of GI, to reduce costs, to improve decision making using spatially enabled analyses, and to expand more efficient service delivery

Five guiding principles underpin the NSDI framework in the U.S.:

1. The most current, complete, and accurate data in any area should be available via the framework.
2. The framework should be user-oriented, i.e., users must be able to easily integrate their own data with framework data and also to provide feedback and corrections to the national framework data.
3. As the NSDI framework data are a public, national resource, access should be at the lowest possible cost and without restrictions on use, dissemination, or reuse.
4. GI production and maintenance costs should be reduced by removing duplication of effort across different GI communities.
5. The framework is based on the principle of wide cooperation, created from the combined efforts of many participants at all levels within the framework, i.e., in design, development, and contributing data.

The four major components of the U.S. NSDI framework are information content, technical context, operational context, and business context. Information content refers to the data in the framework, comprising seven main themes of the most commonly used GI. Technical context includes any technology required to build and operate the framework. Operational context describes the framework's operating environment, and business context addresses the conditions required to ensure the usability of framework data, including business models and identification and promulgation of best practice.

While the 1994 executive order set the policy for the U.S. NSDI, implementation rules were promulgated via the Office of Management and Budget (OMB, 2002) Circular A-16, revised. This document revised an earlier 1990 circular and incorporated Executive Order 12906. Thus, OMB Circular A-16 became one of the main implementing instruments to enact the new U.S. NSDI policy. Yet the 1994 executive order was not the first SDI initiative in America, although it was the first national directive relating to SDI issues. The Mapping Science Committee of the National Research Council had produced a series of reports, from as early as 1990 (NRC MSC, 1990), investigating, among other things, the spatial data needs for a "national mapping program" and the benefits that might accrue. Research completed in 1993 and published in 1994 (before the executive order was issued) had already concluded that the successful creation of the foundation data sets needed to support an NSDI (NRC MSC, 1990) required strong future partnerships not only within federal government, but across all levels of government and with industry (NRC MSC, 1994). Their report advocated "shared responsibilities … shared commitment … shared benefits … shared control" and proposed that the benefits of spatial data partnerships should be evaluated "for the entire national community of spatial data users, not merely for the agencies participating in the partnership" (NRC MSC, 1994, p. 2).

The theme that an NSDI involved more stakeholders than just federal or central government agencies was to reappear more than a decade later with the proposal for The National Map (TNM) program. TNM is the product of "a consortium of Federal, State, and local partners who provide geospatial data to enhance America's ability to access, integrate, and apply geospatial data at global, national, and local scales" (USGS, 2007). It is a partnership effort among the National Geospatial Programs Office (NGPO) of USGS, the National States Geographic Information Council (NSGIC), and the National Association of Counties (NACo).

TNM is expected to help create a better, more comprehensive, more up-to-date national GI resource than had been achieved by 2004 solely within the framework of the NSDI itself, as originally promulgated to federal agencies and based mainly on standards and clearinghouses for all federal GI resources (Lukas, 2004). TNM can be considered a new policy instrument to help achieve the original goals of the U.S. NSDI, as a result of continuing evaluation of the success or failure of prior mechanisms, i.e., entering a new cycle in the policy process cycle (PPC) model defined in section 6.1.1.

In the foreword to a special issue of the *Photogrammetric Engineering and Remote Sensing* journal on The National Map, one of the proposed responses (policy instruments) addressing this weakness in the U.S. NSDI, Ogrosky (2003) summed up the situation as:

> [It is] increasingly being recognized ... that our traditional ways of acquiring, maintaining, archiving, disseminating, and using geographic information must change in response to resource limitations, increasingly sophisticated requirements, the revision of government and private sector roles, and the availability of powerful tools for mapping and analysis.

According to Charles Groat in 2003, then USGS director:

> An important detail in the United States is that we are working together to build a national map, we recognize that in many cases, if not most, higher resolution and more current data exist at the State and local levels (Groat, 2003, p. 4).

The nontechnological, organizational, and information culture issues regarding U.S. NSDI that were still being encountered a decade following the NSDI executive order were expressed by Kelmelis et al. (2003):

> One of the major challenges is to develop new ways to facilitate partnerships of the willing to make the geographic information available, accessible, and applicable. This goes beyond using current technology and organizational relationships.

The cost–benefit of TNM was investigated in 2004 (Halsing et al., 2004, p. 2) and will be discussed later in this chapter in relation to types of cost–benefit analyses (CBAs) that can be performed for SDIs. Continued evolution of the U.S. NSDI is being guided by a Future Directions Planning Team within FGDC (FGDC, 2004) and includes specific activities focusing on the 50 U.S. states' contributions to a national GI resource (FGDC, 2007b).

6.2.3   *Pan-European SDI developments*

At the regional (multinational) level, the most advanced SDI initiative is that promulgated by the European Union (EU), throughout the 27 member states of the EU, set out in the Infrastructure for Spatial Information in Europe (INSPIRE) directive, which came into force throughout the EU on May 15, 2007 (EU, 2007). The legal directive merely sets out the main principles and goals, while separate implementing rules are created in the five main areas covered by the directive. These are for metadata specifications, data specifications, network services of various types, data sharing and monitoring, and reporting implementation of the directive. In the coming years, EU member states must enact national legislation that recognizes the main articles of the directive, as well as a set of implementing rules that enact the directive. As is usual for EU directives, the practical implementation rules are defined separately from the legal directive itself, and may change over time as circumstances change, for example, due to technological change.

Just as the national initiatives in our examples from the U.K. and the U.S. have taken more than a decade to implement even partially, and are still evolving, so too is the case for the European regional SDI. Work began on the main policy visions and strategy development early in 1995, resulting in a legal directive in 2007, for which many of the implementing rules are not required to be in place — and enforced — until 2013 or 2014. Obviously, creating SDIs takes a long time. During the consultation period from 1995 to 1999, relating to the European SDI initiative, then dubbed GI2000, the European Geographic Information Infrastructure (EGII) was loosely defined as encompassing the broad policy, organizational, technical, and financial arrangements necessary to support increased access to European GI. By 1998, a more formal definition had been accepted, which was

> a stable, European-wide set of agreed rules, standards, procedures, guidelines and incentives for creating, collecting, exchanging and using geographic information, building upon and where necessary supplementing, existing Information Society frameworks. The aim should be to create a competitive, plentiful, rich and differentiated supply of European geographic information that is easily identifiable, easily accessible and usable (European Commission, 1998).

The policy framework within GI2000 was expected to address "the political and technical issues of lowering the cost of collecting, disseminating and using GI throughout Europe, thereby improving the functioning of the internal market. It should take into account the wider objectives of public policy, in particular that of ensuring that fundamental rights to privacy are fully respected" (European Commission, 1998). The GI2000 initiative faltered late in 1999 due to political upheavals within the European Commission. The main initiatives and much practical SDI preparatory work continued via a series of EU-funded projects until the concept was renewed in May 2001, resurfacing as the Environmental-European SDI (E-ESDI) within the EC's Directorate General for Environment, to support future pan-European work relating to environmental actions. This resulted, in December 2001, with an action plan to implement the E-ESDI as the first sectoral component of a wider, more generic ESDI (European Commission, 2001). E-ESDI faded from view relatively quickly, being subsumed into the wider INSPIRE initiative that led to the May 2007 legal directive of the same name. One of the main reasons put forward for the eventual success of INSPIRE vs. GI2000 was the direct, high-level political support for the pan-European SDI concept demonstrated in a joint memorandum of understanding in April 2002, signed by the three EU commissioners responsible for Environment (European Commission, 2002a), the Joint Research Centre, and the European Statistical Office (Eurostat). The three commission directorates general whose duties fall under these commissioners continue today with the implementation aspects of INSPIRE, the European SDI.

Between 2002 and November 2006, intensive consultation across Europe resulted in the final agreed-upon text for INSPIRE. During these four years (building of prior project work was completed between mid-1999 and mid-2002), hundreds of experts were involved in investigations of the data needs for a pan-European SDI, the implementation cost ramifications, the potential benefits, impact analyses, and practical issues such as standards for metadata and data, how data would be delivered to users, access principles (and cost regimes), etc. While advances in technology and especially in interoperability standards, tools, and techniques were removing many of the previously identified technical barriers, policy issues relating to access principles, use and exploitation, and charging regimes continued to hinder adoption of an agreed-upon text. The situation was confused by other legal directives enacted prior to INSPIRE that covered access to and use of environmental data, reporting requirements (using spatial data and GIS tools) for the Water Framework Directive (European Commission, 2000), and Re-Use of Public Sector Information generally, 80% of which is proclaimed to be spatial in nature (European Commission, 2002b).

6.2.4   Policy role in other SDI definitions

A decade ago, at the Second Global SDI Conference in 1997, the multinational GSDI Steering Group defined the Global Spatial Data Infrastructure (GSDI) as "policies, organisational remits, data, technologies, standards, delivery mechanisms, and financial and human resources necessary to ensure that those working at the global and regional scale are not impeded in meeting their objectives." The policy role is recognized in the GSDI's "SDI Cookbook," which defines SDI as (Nebert, 2000)

> the relevant base collection of technologies, policies and institutional arrangements that facilitate the availability of and access to spatial data. The SDI provides a basis for spatial data discovery, evaluation, and application for users and providers within all levels of

government, the commercial sector, the non-profit sector, academia and by citizens in general.

The Australian and New Zealand Land Information Council defines the
Australian SDI (ASDI) as "a national framework for linking users with providers of spatial information. The ASDI comprises the people, policies and
technologies necessary to enable the use of spatially referenced data through
all levels of government, the private sector, non-profit organisations and
academia" (ANZLIC, 2007). The ASDI was originally conceived as comprising four core components: an institutional framework, technical standards,
fundamental data sets, and clearinghouse networks. Within this overall
structure, the institutional framework defines the policy and administrative
arrangements for building, maintaining, accessing, and applying the standards and data sets.

The Canadians become a bit more precise in defining their national SDI
— the Canadian Geospatial Data Infrastructure — called GeoConnections,
which has five main policy areas:

1. Policy for accessing data
2. Policy to establish a framework of data to enable easier integration to
   aid decision making and develop new information products
3. Standards policy to ensure that Canadian information matches international standards
4. Partnerships policy to encourage and ensure collaboration at various levels
   of government and with the private sector and the academic community
5. Supportive policy at all levels of government to accelerate private sector
   commercialization of geospatial information, and to develop e-commerce and integrated technologies and services

In the Asia-Pacific region, the Permanent Committee on GIS Infrastructure for Asia and the Pacific (PCGIAP) has a vision for an Asia-Pacific Spatial
Data Infrastructure (APSDI) that includes a network of databases distributed
throughout the region to provide the fundamental data needed across the
region to achieve its economic, social, human resources development, and
environmental objectives. Two key objectives of the information policy in
the APSDI are:

• To increase the ability to share data, which will then reduce duplication
  of resources and facilitate data integration across sectors, users, and
  national boundaries
• To provide better data for better decision making and to help expand
  market potential for geographic information

The APSDI information policy establishes a set of principles for responsible management of regional GI and commits all countries in the region

to cooperate in the implementation of the APSDI to implement the principles. Unlike the European regional SDI INSPIRE directive, there is no policy enforcement mechanism applicable to the APSDI. Thus, the 55 countries of the region who belong to the PCGIAP may take up the principles or not, as time and resources allow.

In Africa, the UN Economic Commission for Africa (UNECA), the Global Spatial Data Infrastructure (GSDI) Association, and EIS-Africa, in collaboration with the International Institute for Geoinformation Science and Earth Observation (ITC) in the Netherlands, created a national SDI implementation guide (SDI Africa, 2004). The objective of compiling this handbook was to assist African countries to improve the management of their geospatial data resources in a way that effectively supports decision making by governments and ensures the participation of the entire society in the process.

Another study published in 2001 by the UNECA's Development Information Services Division, Geoinformation Unit, looked at the importance of SDI both nationally and regionally in Africa. The report (UNECA, 2001) identifies the main components for an SDI (similar to those in most SDI vision and strategy documents), then examines a range of issues related to implementation of SDI nationally and for the region, including policy considerations. The paper identifies "a need for a geoinformation policy, within an overall information management policy," and provides a "Model Policy and Institutional Framework for SDI" in an appendix to the report (p. 12). The model policy comprises a statement of vision, principles, and three major policy guidelines, including:

1. A national geoinformation framework should be created that comprises:
   - National geoinformation with broad representation from society
   - Improved communication between stakeholders, including institutional producers and users of data
   - Use of appropriate ICT for improved access to GI resources by all stakeholders
   - Creation and maintenance of fundamental (core, reference) geospatial data sets, and the metadata systems necessary for their discovery
   - Increasing the number of skilled personnel to maintain the SDI framework and data sets and the level of knowledge and skills in the community of stakeholders to make the most effective use of the data sets
   - Developing and implementing appropriate pricing mechanisms for data usage
2. Publicly funded development plans should include details of the geoinformation requirements needed by the plan.
3. All public project proposals dealing with infrastructure development and maintenance, environmental and natural resources management, and spatial facilities shall include information budgets.

6.2.5   Summary of policy roles in SDI formation

From the previous section, we see that a range of different SDI definitions emerge, centered on the practices of geographic information interoperability shared among a range of public and commercial players. The settings in these definitions are national and regional, although certain transboundary and transsectoral initiatives have been established. As to the types of policy identified, within the typology described earlier, comprising distributive, redistributive, regulatory, constituency-based, and governance-based policies, we see that SDI policies do not fit neatly into any one category.

Key to all SDI visions and one of the main policy statements found in all SDI strategies is the importance of policy for access to information. Access is defined in different ways and at different levels of functionality, ranging from relatively simple metadata access, so that a potential user can find a data resource of possible interest, to full download capability with no restrictions on use or reuse, including for commercial exploitation. Access issues include technology to enable access, standards for both data and metadata, and pricing or charging for access, whether for own use or commercial exploitation. Chapters 3 and 4 covered the pricing and charging issues quite well, and Chapter 5 (section 5.4) presented a comprehensive overview of many of the political issues surrounding SDI formations, as empowering or controlling or legitimizing infrastructures.

While most SDI policies and strategies actively promote free access to metadata, preferably published by electronic means via the Internet, as soon as one moves up the information functionality scale even to the level only of viewing data (with no download or printing capabilities), policies begin to diverge. Due to the lack of full SDI vision or strategy implementation in almost any country and the resulting lack of postimplementation evaluation, it is difficult to determine what policies and strategies have proven most effective in achieving broad access, use, and reuse goals for GI — voluntary, best practice, official recommendation, or legal requirement. What is apparent from the previous sections, and is the focus of the following section, is that policy implementation strategy is nearly as important as the policy formulation process itself, since the most beneficial policy in the world can be thwarted by poor implementation of the actions needed to support that policy.

6.3   Implementing SDI policy

In this section we look at how SDI policy is implemented, typically using an implementation plan conveyed in a formal information strategy developed to implement a stated information policy. However, remember the close link between public sector geographic information (PSGI) and public sector information (PSI) policies inherent in national information infrastructure (NII) and e-government initiatives. These links result in some goals of SDI policy and strategy being achieved vicariously, for example, courtesy of national and global standards for data representation, metadata, national information access portals, and digital rights management technologies, e.g., click-use licenses for online access.

How does information policy differ from information strategy? At the simplest conceptual level, policies are set and strategies are performed in implementing the policy. However, like policy, strategy can also be a thing (product) and a process. Strategy as a product is typically an expression of a logical and interconnected set of actions, defined in a strategy document, containing an implementation plan, etc. Strategy as a process is the implementation of the plan. Orna (1999) defines information strategy as "the detailed expression of information policy in terms of objectives, targets, and actions to achieve them, for a defined period ahead." The strategy provides the operational framework for managing information and implementing the policy goals.

6.3.1   Policy vs. strategy

Policies define frameworks within which certain goals are expected to be achieved, whether these are data access policies, information exploitation policies, or data privacy policies. An example of a typical policy statement is a legal directive from the European Union, an act of Congress in the U.S., or a decree from some similar national ruling body. Of course, policies and policy statements also exist much lower down the organizational scale, right down to the level of policies set by individuals, e.g., "It is not my policy to watch television on Friday nights." But all policies have four inherent components:

1. A rationale for why the policy is needed
2. An expression of principles underpinning the policy
3. A statement of the goals or objectives to be achieved by the policy
4. Reference to some strategy or action plan that will implement the policy

Note that policy statements need not, and typically do not, include an implementation plan directly, but underpin and justify the legitimacy for a strategic plan and its execution, which may involve considerable cost, organizational change, or even new legislation. The question arises as to whether we should consider policy as the implementation tool for strategy or vice versa.

The *Cambridge Advanced Learner's Dictionary* defines strategy as "a detailed plan for achieving success in situations such as war, politics, business, industry or sport, or the skill of planning for such situations." Merriam and Webster's Online Dictionary includes several definitions of strategy, of which the most appropriate for our discussion is "a careful plan or method, a clever stratagem (a cleverly contrived trick or scheme for gaining an end) or the art of devising or employing plans or stratagems toward achieving a goal." The second definition includes the concept of strategy as both a thing (product) and a process (implementing plans), just as policy could be product or process. Strategies are also defined within, or comprise, frameworks within which their various actions are implemented.

Since policies and strategies are defined for different reasons, comparing one to another is basically an analytical exercise to see if the strategy proposed will achieve the goals of the policy. Unfortunately, comparing a policy framework to a strategy framework is complicated by the purpose of strategy (to achieve specific goals using different measures or instruments) vs. policy (set long-term goals under some guiding principles). One can question which comes first, strategy or policy? Looking at SDI developments in some nations and regions, including in Europe, it seems that because strategy defines near- and long-term goals that are to be implemented as a result of policy, strategic thinking may precede policy formulation, or at the very least progresses in parallel. In practice, the latter is more likely as, during the policy creation process steps defined by the PPC model, policy makers must take into account the ramifications vis-à-vis strategies to implement the policies being developed to achieve the stated goals. This takes place mainly in steps 2 and 3, where policy instruments are proposed, along with alternatives. Different strategies represent different alternatives using different implementation measures (instruments).

What are the real differences, then, between policies and strategies? Blakemore argued earlier in this book that strategy is a dynamic process. In the context of e-government, strategies appeared to be little more than central plans (Blakemore and Dutton, 2003), comprising a set of promises to citizens on a range of issues for which promise fulfillment was more important than strategy monitoring or review. We seem to have a chicken and egg situation — policy is designed to achieve strategic goals, but setting strategic goals depends upon outcomes expected from implementing policies. Any confusion may lie in terminology, in that strategic goals are not the same thing as strategy. The former are legitimate components of a policy; in fact, they are at the heart of policy making. The latter refers to the plan of action undertaken to achieve the strategic goals. Some confusion still remains, in that strategies can cover varying time spans, i.e., near-term strategy vs. future or long-term strategy, and the cumulative achievement of goals set within strategies should result in ultimately achieving the strategic goals of the policy.

Once a policy has been agreed upon, along with an accompanying implementing strategy (which may change over time, following set review periods), the question arises on how to monitor achievement of policy goals vs. strategy goals — or is there no difference? Strategies tend to be specific, set for prescribed periods, may change at the end of the review period, and may have many components all leading to achieving a single overriding policy goal, e.g., increasing the size of the European information market. Yet within any one strategy, there can be many individual goals for the different implementation measures that the strategy prescribes, the success or failure of which can be used to judge the success or failure of the strategy. Remember that it is logically feasible to reach each and every goal set within a specific strategy and yet not achieve the overall policy goals if, for example, the strategies were ill-defined in the first place.

For example, a typical policy goal in most SDIs is to increase stakeholder and potential new user awareness of the existence of GI from which they may benefit. Typical GI implementation strategies nearly always include creating some form of geoportal via which data holders can publicize their holdings and potential users can find them. So one goal of the strategy is creating the portal, and another could be populating the discovery portal with X records by date Y from Z number of organizations. All of this can happen, satisfying multiple strategic goals, and yet the policy goal can remain unmet if no one uses the discovery portal for other legitimate reasons, which may be related to lack of training, lack of bandwidth, lack of appreciation of how to use other people's data sets, etc. There are sadly more than a few such cases in existence today in national SDI initiatives.

One can also question the value of developing and implementing a strategy that lacks enforceability within the policy framework via rules and regulations underpinned by an accepted policy statement, which may itself take the form of legislation. In the U.K., both the national e-government information discovery (metadata) framework (e-GMF) and the underpinning metadata standard (e-GMS) (Cabinet Office, 2006), part of the wider e-government interoperability framework (e-GIF) (Cabinet Office, 2005), were widely promulgated to local government, yet there was no requirement that they actually create PSI discovery portals using this framework or standard, so most did not. The U.K.'s national geoportal, GIgateway, was developed at considerable cost, including two revisions of the standards required to eventually meet the requirements of both e-GMS and the GI industry's ISO 19115 GI metadata standards. However, again there was no requirement that local government — or anyone else for that matter — actually populate or use the gateway, which remained sadly underpopulated some years following its creation. In a 2005 review of the national agreement that paid for maintaining the U.K.'s GI discovery portal (which expired in March 2006), it was recognized that GIgateway was a "potentially powerful tool" but that after 3 years of operation it "did not currently have the critical mass of users to encourage wider uptake." The report also expressed concerns, including "the scope and relevance of metadata available," and noted that "a number of other similar metadata services are operating, focusing on the specific needs of their target user base" (ODPM, 2005). While performance of GIgateway as the U.K.'s national geoportal may have improved by the time this book is in print, the lesson learned is that a valid policy aimed at expanding use of national GI resources may not always be fulfilled if the strategy is found lacking for whatever reason.

The EU's INSPIRE directive is an attempt to force national governments across the EU to create GI metadata portals to a common standard (based on ISO 19115), over a number of years, for 34 different data themes, for any data that contain a spatial component and for which collection and use are legally mandated. Finally, what about the ability of policy makers to forecast the effects that rapidly shifting information industry environments may have on policy goals and even specific strategies? The rapid advances in online geospatial product and service capabilities and offerings, many of which have a direct impact on SDI policies and strategies, are a good example. A strategic policy goal of increasing access to large volumes of public sector GI via a strategy incorporating the latest ICT tools available in 2007 may be more easily achieved — or completely thwarted — by changes in the technological or even legal environment, e.g., arrival of a new IPR paradigm or advanced digital rights management tools and techniques.

How do governments accommodate short-term or medium-term shifts in information policy, and do such shifts occur for policy more frequently than for strategy? Can sometimes subtle shifts in policy caused by implementing measures used in a strategy negate or lessen the intended impact on the policy's strategic goals that were defined by higher-level actors, e.g., politicians, trade bodies, or heads of government agencies? How does one attempt to foresee or measure this type of impact? These are but a few of the many issues facing both policy makers and decision makers, which unfortunately we do not have room to more fully explore in this chapter.

6.3.2 Policy conflict and harmonization

What about conflicting information policies? For example, in the U.K., citizens, government, and businesses experience the Office of Public Sector Information's (OPSI) strong promotion of open-access and reuse policies for all PSI (including PSGI) vs. the existence of trading funds, which charge for use of all (or most) of their information resources, typically via licensing. In this situation, two conflicting issues emerge when trying to ensure widespread use of scientific data collected for environmental and global change monitoring and research (Longhorn, 2002). First is the desire that such data be as widely shared and used as possible via a full and open policy, which may depend upon uncertain funding by central government, which varies over time, resulting in uneven data coverage, quality, and timeliness. Second is the desire of some governments and agencies to recover costs for data collection, processing, and dissemination operations, partly to ensure higher data quality and continuity of data collection without dependence upon central government funding. Which of these is the better policy was explored in Chapter 4, without any definite answer being possible, because the answer also depends upon a number of circumstances particular to each nation or even period within a nation's information society development.

Consider data protection (personal privacy) policy relating to location-based services (radio frequency identification (RFID) tracking, emergency location from wireless 911 calls in the U.S., georeferencing of CCTV footage, etc.) vs. personal privacy, personal liberty, and personal protection. If you are lost and injured on a hiking trip, you will gladly accept the help of a system that reports your location to an emergency rescue team automatically based on your cell phone location. Yet you are not at all pleased if your boss finds out that you were not at home sick yesterday (as you claimed) when a similar commercial service reports your sick call as coming from the location of your known holiday home on the lake or the local baseball park or cricket ground. Police use of georeferenced CCTV camera footage to track, model behavior, and finally capture a gang of thugs who have been beating up elderly people in shopping malls across Chicago, London, or Tokyo is welcomed by the citizens of those countries. These are the same citizens who then complain of excessive spying on their own innocent movements, which is a by-product of trying to implement personal security via increased CCTV camera coverage.

Conflicting information policies are the reason for initiatives to seek harmonized policies across sectors, across borders, within government, and among government, business, and citizens. Weiss and Backlund (1997, p. 309) noted the conflict with regard to cross-border meteorological data that occurs within the context of a long-established international framework for the production and distribution of such information via agreements within the World Meteorological Organization. "The conflict between the public good/private enterprise partnership arrangements followed in the U.S., exemplified by the diversity principle ... and the efforts of some government entities to restrict the flow of information for quasi-commercial purposes is threatening the traditional framework of open and unrestricted exchange of weather related data."

6.4   SDI cost–benefit issues

What is the impact on policy implementation of resource requirements, e.g., finance for capacity building or education and training to create more aware users? Why support a policy that cannot be implemented due to lack of adequate resources or because of other barriers? What are such other barriers? In this section we examine the extent to which the value (benefits) of GI to a national economy can be quantified — or not — and whether the often high (expected) costs to create spatial data infrastructure (SDI) can be justified. The fact that SDI is part of the more generic national information infrastructure (NII) in a country (or even within a very large organization) adds complexity to the analysis and justification. This is because many of the components of an SDI are also included in NII or e-government initiatives, so where should the cost be allocated, to the NII or the SDI?

In this chapter we will revisit and summarize some GI and GIS cost–benefit studies conducted in the past 15 years, up to the present time (early 2007), for different countries globally and for the EU as a region. It is useful to examine similarities and dissimilarities in assumptions, approach, methodology, goals, and terms of reference of the studies. Along the way, we address a number of important questions. One major question still facing many SDI initiatives is whether traditional cost–benefit analysis (CBA) methodologies can be used effectively for information infrastructures as opposed to individual projects, for which such methodologies were originally developed and are traditionally used. Can a CBA methodology that focuses on the value of GIS be used to examine the value of GI or the value of SDI? Are the CBA methodologies from past studies still applicable today, and if so, where, when, and how? Can the results of any one study be applied generally to the value of the GI debate globally? Can you characterize the assumptions and methodologies in a study in a formal way to help determine if the study results can be applied elsewhere? If not, how and why should decision makers rely on the predictive results of the preexisting value of GI studies carried out under different assumptions and circumstances? What alternative methods exist for examining the value of geospatial data to an SDI, e.g., simulation programs that implement economic models for different scenarios in creating an SDI or other predictive software tools.

6.4.1   *Historical SDI CBA results*

Many cost–benefit studies have been conducted in relation to geospatial information system (GIS) projects and technology, beginning as far back as the mid-1980s. Far fewer have been conducted looking specifically at quantifiable benefits for implementing entire infrastructures, such as spatial data infrastructure (SDI). Table 6.1 shows a range of typical studies over time, geography, nationality, sector, and diversity of type of study. Some of these studies investigated cost–benefit for only single industry or government sectors or agencies, or types of spatial data technology or applications. Others covered a wider range of sectors and regions, from national to transnational. Some studies considered only public sector GI, while others tried to factor in the impact of private industry on SDI strategies and the impact of government SDI policies and strategies on private industry. A few studies looked only at quantifiable monetary revenue as the benefit, or at savings in labor time, to which a cost savings was then attributed. Others attempted to assign monetary values to more qualitative benefits accruing to society generally, i.e., to government for efficiency savings, new services not previously available, etc., or to businesses in creating new, more competitive services, and to citizens for security, convenience, time savings, etc.

Prior to expending the time and money on conducting a CBA, one question that requires an early answer from the funding policy makers and allied decision makers is what level of imprecision is acceptable in the results without automatically negating a decision to provide funding? Is it sufficient to demonstrate that there is a reasonable expectation that the benefits will outweigh the costs, or must very specific targets be met, i.e., return on investment (ROI) must be at least 20% within 3 years, or the benefit–cost ratio must equal or exceed 4:1? Reaching early agreement on this will better inform

Table 6.1   Benefit:Cost Ratios from Prior CBA Studies

Date	Organization	Country	Type of Study	Benefit: Cost
1990	New South Wales state	Australia	Economic aspects of digital mapping	2:1 to 9:1
1990	Western Australia Department of Land Administration	Australia	Land information program	5.9:1
1991	Office of Information Technology of South Australia	Australia	GI in the public sector	2.9:1 to 5.8:1
1992	AUSLIG	Australia	Economic and social benefits of public interest program	3.8:1
1992	Department of Defense	Australia	Economic benefits of hydrographic programs	2.7:1
1993	Government of Victoria	Australia	Strategic framework for GIS development	5.5:1
1994	Australian Bureau of Agricultural and Resource Economics	Australia	Economic analysis of remote sensing for land management	[1]
1995	ANZLIC	Australia/NZ	Australian land and geographic data benefits study	4:1
1996	Coopers and Lybrand for OSGB	U.K.	Economics of collecting, disseminating, and integrating government GI	N/A
1998	U.S. Department of Agriculture	U.S.	ROI for GIS projects from agency-wide business process re-engineering study	$168 million in savings/year
1999	Department of Land and Water Conservation, NSW	Australia	Business case for community access to natural resources information (1999–2003)	1.82:1 average
1999	OXERA for OSGB	U.K.	Economic contribution of OSGB	[2]

Table 6.1 Benefit:Cost Ratios from Prior CBA Studies (Continued)

Date	Organization	Country	Type of Study	Benefit: Cost
2000	PIRA International (U.S.)	EU-wide	Commercial exploitation of Europe's public sector information	[3]
2000	Center for International Economics, Sydney (for GSDI)	Global	Describes preferred methodology for preparing business case for SDI	N/A
2001	Baltimore County (Maryland) Office of Information Technology	U.S. (local government)	10-year forecast CBA for savings across local government departments using GIS and geodata	IRR = 64–168%
2002	Austrian Federal Ministry of Economics and Labor	Austria and Europe	Economic analysis of CBA for Austrian cadastral GI	23:1[4]
2003	Environment Agency U.K. and University of Sheffield, U.K.	EU-wide	Contribution to the extended impact assessment for INSPIRE	4.4:1 to 8.9:1
2004	European Commission INSPIRE	EU-wide	Extended impact assessment for INSPIRE	5.4:1 to 12.4:1
2004	U.S. Geological Survey U.S. Department of Interior	U.S.	Determined net present value (NPV) of U.S. National Map program over 30 years	$2 billion benefit
2005	Booz Allen Hamilton (U.S.)	U.S.	Geospatial interoperability ROI study	ROI = 26.2%

Note:

[1] Remote sensing returned net gain of AUS$1.5 million and AUS$66 million in monitoring trees and fertilizer use.

[2] OXERA reported estimated value to U.K. economy of £100 billion (£100,000 million) from GI maintained by Ordnance Survey GB at an annual cost of around £100 million.

[3] Economic potential to society of wider use of PSI, of which GI played a major part (over 50% of total PSI value).

[4] The Austrian analysis includes tax revenues in the benefits to the state, as well as registration fees; this is more a monetary revenue:cost ratio than the CBAs reported in other studies.

N/A = No specific figures are stated or the studies looked mainly at nonquantifiable, qualitative benefits, so figures are not available or looked at benefits of GIS, and thus are not applicable.

those who conduct the CBA, so that appropriate methodology and effort can be expended, not wasted.

As with many cost–benefit studies, virtually every study referenced in Table 6.1 had much greater difficulty quantifying benefit compared to cost. The studies typically include benefits from:

- Reducing duplication of data collection costs
- Creating fundamental (core or base) reference spatial databases to an agreed standard so that new products and services can be developed more cheaply and more quickly
- Providing wider access to data sets, including discovery via standard metadata and publication of metadata and data via interoperable geoportals
- Commercialization of government agency data activities, including potential disbenefits
- Better coordination of spatial data collection and publishing regimes across all government departments
- Outsourcing of some specialized tasks
- Efficiency gains from wider access to better-quality data both internally to an organization and across organizations and disciplines, usually quantified by estimates of labor cost savings
- Additional positive impact on other projects or infrastructures made possible by rapid access to GI and metadata
- Benefits to society, which are seldom quantifiable financially, but may be even more important than many directly attributable financial benefits, e.g., better policy making, implementation, and monitoring
- Other macroeconomic benefits of a generic nature that may apply to all forms of information, not just spatial data, and are thus even more difficult to determine in relation solely to SDI

Benefits were examined from different viewpoints in the different studies, i.e., impact on a specific project, impact on an agency, impact on users, impact on a wider community or on society as a whole. While significant benefits may accumulate for a wide community or section of society, such benefits are also the most difficult to quantify in financial terms. If implementation of an SDI supports other projects or infrastructures, how can one accurately assign a portion of the total financial or societal benefit derived from that project or initiative to the SDI component? Funding agencies worry about double accounting for such benefits, i.e., claiming the benefit as accruing to both the SDI and the project or infrastructure supported by the SDI.

A comment often encountered in the studies reviewed is that "many of the most valuable benefits are in areas of improved services and performance rather than direct cost savings. These improvements are particularly difficult to quantify" (Montgomery County Council, 1999, p. 3). Also "Dutch study [found] for every euro invested in SDI after a few years yields 10 euro of benefits ... (and an) extended impact assessment for INSPIRE ... showed significant benefits. But the underlying calculations of these studies are predominantly based on estimated and extrapolated parameters. The benefits are not proven" (Bregt and Crompvoets, 2005, p. 5).

Most CBA studies presume that cost figures are better defined than benefits. Yet costs are not always easily attributable, especially if they are additional (marginal) costs, associated with a project or other infrastructure that is being developed in any case. For example, to whom does one attribute the cost for implementing user identification, authentication, certification, and verification procedures for access to a government information system generally, as opposed to the same cost to access a national government GI portal — to a general e-government program or to the SDI initiative? Costs are also handled differently, depending upon who the study is conducted for, by whom, and why. The INSPIRE impact assessment looked only at incremental costs and investment requirements for creating the pan-European SDI, i.e., those costs over and above what would be expended in any case by national governments in collecting and disseminating spatial data, even without an INSPIRE directive (Craglia, 2003).

Typical cost categories include those shown below, which may be relatively easy to determine for a single project, but are much more complicated for an entire infrastructure:

- Additional hardware and GIS software or related IT infrastructure costs, especially ongoing maintenance
- Additional personnel and training costs, including over long periods of time as staff change posts or the information infrastructure evolves, placing new demands on data custodians
- Geospatial data collection costs or purchase/licensing costs
- Database updating, maintenance, and dissemination, over various periods of time, ranging from project lengths of 3 to 4 years up to decades in the case of whole infrastructures

Other issues that arise in relation to cost–benefit include:

- Timing of investments to implement spatial data infrastructure for different types and levels of spatial data, e.g., core data vs. noncore, resolution of granularity of data, level of metadata (discovery or exploitation)
- Time for benefits to begin to appear in a quantifiably measurable way and definition of the metric to be used
- Risk assessments, e.g., impact of delays in taking action, technology impacts, lack of engagement of all stakeholders, disharmonized or unsynchronized implementation of SDI framework elements
- Assessment of potential disbenefits arising from different implementation policies, e.g., data access policy, data exploitation policy, legal issues

Our research examined alternative approaches used in conducting typical cost–benefit studies, specifically the traditional cost–benefit analysis (CBA) vs. cost-effectiveness vs. hybrid methodologies combining traditional CBA approaches with Delphi techniques, i.e., interrogating experts in relevant fields. The different approaches yield different types of results, especially with regard to precision of monetary figures needed for budget allocation decisions on how much to invest, and when, in different SDI implementation regimes. The INSPIRE Extended Impact Assessment Report (Craglia, 2003, p. 28) noted that "the challenge [in completing the impact assessment] was the almost complete absence of previous studies containing quantitative information on the costs and benefits of introducing infrastructures for spatial data." This sentiment is shared in numerous other publications from senior and well-respected experts from the GI/GIS industry, academia, and government. Several of the studies listed in Table 6.1 separately note that very few CBA studies (if any) have been conducted postimplementation to look at actual benefits achieved, even at the project level. The impression is that once the decision has been made to proceed, there is little interest in expending yet more resources at a later time to try to accurately measure the benefits, except for those financial benefits that are reported naturally as part of annual accounting, auditing, or other organizational review practices.

6.4.2 SDI CBA methodologies

Cost–benefit methodologies include the traditional financial cost–benefit analysis (CBA), cost-effectiveness analysis (CEA), and hybrid methodologies, such as multicriteria analysis (MCA), the value measuring methodology (VMM), and simulation modeling. These latter methods combine some traditional financial CBA elements with nonmonetized cost–benefit impact assessments either using Delphi techniques, i.e., interrogating experts in relevant fields or stakeholders, or by running extended computer-based simulations of multiple scenarios. The different approaches yield different types of results, especially with regard to precision of monetary figures needed for budget allocation decisions on how much to invest and when. All these methods can be applied using different levels of detail depending upon the type of project or program under evaluation and the planned scope of the evaluation, keeping in mind the criteria of the funding body. If public money is involved, a social perspective is likely to be required that is absent from project assessment in the private/commercial sector. In this case, the evaluation method is sometimes called social cost–benefit analysis (SCBA) or social multicriteria analysis (SMCA).

The analysis methodologies are applied on different levels, for example:

- At the policy level, CBAs are used to analyze the impact of different policy options, for example, in INSPIRE, prioritizing the types of data (e.g., across the three annexes) to be included in data harmonization measures required by EU member state governments or the level of harmonization requested of different data types, where one level is more expensive to achieve than another.

- At the strategy level, analyses examine the cost–benefit of implementing specific implementation strategy components for specific economic sectors or specific data themes, based on a readily identifiable or expected benefit or outcome.
- At the financial or budgetary level, analyses might examine investment alternatives to see which data harmonization and interoperability actions within an SDI yield the highest monetary return within fixed budgetary parameters.

6.4.2.1   Traditional cost–benefit analysis (CBA)

The traditional CBA is widely used, covers as many factors as the investigators think are feasible to examine for both costs and benefits, and uses these statistics vigorously, but tends to be best at evaluating specific projects, vs. larger infrastructures. The goal of a CBA is to determine the net economic gain from a particular investment (program or project), typically in monetary terms, across all phases of a project or program, on an economy-wide basis while attempting to accommodate intangible factors (especially with regard to hard-to-quantify benefits). The costs and benefits are typically adjusted to their present value, so that a net present value (NPV) figure can be computed for the investment, which may then be subjected to sensitivity analyses. Results are presented as either negative or positive NPV or the equivalent as a benefit:cost (B:C) ratio.

Using the traditional CBA methodology, if there is a net gain and this is greater than the gain from an alternative investment opportunity, if sufficient funds are available, and if the quantitative estimates are supported by qualitative evidence, then the investment is made. However, for large infrastructure investments, especially by governments, numerous other factors also come into play. The B:C ratio for investing in program X may be higher than for program Y, but there may be any number of reasons why program Y must be enacted anyway. Sadly (for governments and politicians), government investment decisions are seldom that black and white. The types of costs and benefits examined in the typical CBA have already been introduced in an earlier subsection, so will not be repeated here. Suffice it to say that costs are often grouped into direct costs (fixed and variable, applying directly to the project or program) and indirect costs (part of an organization's general overhead).

Benefits can be grouped into those affecting suppliers (greater efficiency in data collection, dissemination, etc.) and users (value to the user for his or her specific use of the data, i.e., coastal, forestry, or agricultural management). For larger projects, and certainly for most infrastructure programs, there are also wider community (socioeconomic) benefits to be considered.

These can be especially difficult to identify or estimate with any accuracy, without resorting to sophisticated economic modeling. It would thus appear that the traditional CBA approach is not best suited to economic analysis and investment decision making for large infrastructure projects, yet also seems to be the approach that has been used in many of the studies done in the past decade.

6.4.2.2   *Cost-effectiveness analysis methodology*

Cost-effectiveness analysis tries to overcome some of the problems with the traditional CBA by examining benefits using a smaller number of indicators, which act as proxies to reflect overall advantages of implementing the infrastructure. Cost-effectiveness looks at competing ways for creating or maintaining a given capability (in this case, geospatial data collection and provision to or for a large variety and number of stakeholders), using the relative costs of each as a guide to which is best. These costs are calculated just as they would be for the traditional CBA.

However, benefits are measured in terms of the difference in cost between an already established method, e.g., geospatial data provision by paper map, and the next best alternative, e.g., data provision on magnetic media, such as a CD-ROM, or via the Internet for yet another alternative to be examined. Thus, benefit is defined as a cost savings from making the right choice(s) between competing alternatives for any one function or capability in the system, project, program, or infrastructure under consideration.

The method does have its flaws; for example, it assumes that all methods are equal in quality or quantity of output they can achieve, which is patently not the case when choosing between disseminating geospatial data on a hard-copy map and serving it from a database via the Web, on the fly (more current, cheaper, more immediate, etc.). Also, it may be more difficult to identify other types of efficiency gains or wider community benefits, unless special care is taken in the analysis to specifically include these. However, because these latter benefits are often not identifiable or quantifiable, accurately, even in traditional CBA analyses, this is not a reason to reject the cost-effectiveness approach.

The methodology does lend itself to examining creation of information infrastructure at the government level, because the geospatial data at the heart of the infrastructure is already needed, collected, processed, disseminated, and used today anyway. The infrastructure, new or enhanced, is supposed to introduce efficiency gains, as well as (perhaps) some new products and services, but these latter are probably of more importance to businesses than to government agencies (except for those that operate as businesses). It is just such efficiency gains that this methodology can identify.

Cost-effectiveness analysis methodology is best suited to situations where commitment to the overall level of investment funding is already established, to broad guidelines. The value of this method is in determining whether funds already allocated have been or are likely to be efficiently utilized. Thus, for those authorities still wavering as to whether to commit to the often extensive funding needed to create an information infrastructure (whether for geospatial data or other information), this is probably not the best methodology to be used.

6.4.2.3   Multicriteria analysis (MCA)

Multicriteria analysis (MCA) is also referred to as multiobjective decision making, multiobjective decision support system (MODSS), and multicriteria decision aid. MCA helps policy makers make strategic decisions when progress toward multiple objectives cannot be measured in terms of a single criterion alone, such as monetary value used in traditional CBA methodologies* based on economic or financial efficiency criteria, e.g., NPV or ROI. MCA incorporates other types of criteria, such as distributional, equity, ecological, and social value, where alternatives are not based exclusively on monetary values and, in many cases, cannot be assigned monetary proxies. MCA methods can evaluate quantitative or qualitative data, or a combination of the two. Compared to a CBA, MCA enables a more realistic representation of the problem and allows the trade-offs to be more explicit. The interactive nature of the approach enables both the analysts and the decision makers, who could be a number of groups of stakeholders, to learn more about the problem. Although MCA is a structured approach, it is flexible enough to allow the use of value judgment, similar to the value judging that takes place in the benefit analysis of the value measuring methodology (VMM) described later. Finally, MCA is suitable for problems where monetary estimates of the effects are not readily available or knowable.

On the negative side, there is the possibility that preferences will be determined by a single decision maker, without consultation with the community, unless this possibility is recognized at the outset and steps are taken to include all relevant stakeholders — or at least as many as possible, a minimum being a good representative sample of stakeholders. Even though MCA does not require quantitative or monetary data, the information requirements to compile the effects table and to derive weights (see section 5.2) can be considerable. Improperly applied, MCA has the potential to become opaque, producing results that cannot be explained easily. This destroys one of the main benefits of the methodology, i.e., the openness and transparency needed to trace back all steps of the decision-making process, for both stakeholders who are affected by the final decision and the decision makers themselves.

Thus, MCA is a decision-making tool for use in complex situations where multiple criteria are involved, and financial/monetary figures may be hard to arrive at or derive, but where a well-structured, transparent, and logical decision-making process is required. It helps address the problem of reaching a general consensus in a multidisciplinary team where members do not always

---

* In traditional CBA methodologies, alternatives are evaluated by performance criteria measured in monetary terms, typically using only quantitative data.

agree on the relative importance of the selected criteria or on ranking the alternatives. In an MCA process, each team member offers his or her own judgments, and thus makes an identifiable contribution to the joint decision, and both qualitative and quantitative aspects of the problem can be considered.

The MCA methodology provides a structured and traceable analysis, copes with large amounts of information, and permits different evaluation factors to be used, making it ideal for discussion within multidisciplinary groups. There are also different ways that MCA can be applied, i.e., different tools have been developed for using the MCA process to examine different types of problems. On the negative side, some consider MCA to be too subjective. However, since many of the assumptions built into the more financially oriented cost–benefit analysis methodologies are equally subjective, at least MCA provides a formal tool and accepted methodology that makes the subjective element transparent to those who participate in the analysis and those who use the results to make decisions. The emphasis must be on selecting appropriate criteria for judging the decision under investigation. For SDI, the criteria may vary from group to group across the wide range of decisions that are needed in implementing something as complex as an information infrastructure.

Many MCA methods (applications and tool sets) are available, because different types of decision making benefit from the general methodology, where the time and resources (human, budgetary) available for the analysis vary, as do the amount and quality of data available for analysis. The skills of the team conducting the analysis will vary, as well as those participating, along with the organizational cultures of the participants. MCA methods can be used to identify a single most preferred option, to rank options, to short-list a limited number of options for subsequent detailed appraisal, or simply to distinguish acceptable from unacceptable possibilities. The very diversity of the MCA methodologies and the ability to separate the decision elements and trace back the decision-making process make MCA ideal for communicating the basis of decisions to a range of decision makers. GeoVMM is a form of MCA, as described later.

6.5   Samples of SDI CBA studies

In this section we present a brief review of some of the more important SDI-related studies produced between the seminal work of Price Waterhouse in 1995 and the latest ROI for using interoperability technology from NASA in 2005. Much of the material presented here was developed by the authors to inform and guide national SDI strategy developments, for example, in Ireland, and for a more global audience, via participation in a special workshop convened by the European Commission's DG Joint Research Center in January 2006 (European Commission, 2006).

6.5.1 Price Waterhouse Australian SDI study (1995)

The Price Waterhouse (PW) Economic Studies and Strategies Unit "Benefits Study" (Price Waterhouse, 1995) for the Australia New Zealand Land Information Council (ANZLIC) built on methodologies and prior studies dating from as early as 1986 (Canada) to 1988 (ACT, Australia), 1990 (Western Australia), 1991 (South Australia), 1992 (AUSLIG — economic and social benefits study), 1993 (Victoria, Australia), 1994 (economic value of remote sensing in Australia), and others from outside Australia. The benefit–cost ratio of approximately 4:1 for geospatial data usage became the beacon to other nations that were considering implementing SDI-type infrastructures, and this ratio is still repeated in SDI strategies, visions, justifications, and other CBAs today.

While many of the previous studies (above) focused on either narrow sectors or themes, often at the state level (not federal), the remit from ANZLIC for the 1995 PW study was specifically national "to determine and prioritise the steps data supplying organisations in Australia should take to maximise potential infrastructure benefits." The study concluded that for every AUS$1 invested in producing land and geographic data, AUS$4 was generated for the economy, amounting to AUS$4.5 billion in the period from 1989 to 1994. The survey conducted by the investigators reported that use of existing infrastructure (in 1994, when the survey was carried out) had saved users "over AUS$5 billion" over the past 5 years alone, during which period data-supplying agencies had invested "close to AUS$1 billion" in their infrastructure.

The study also forecast that government-owned data-supplying agencies would need to grow their data provision budgets "in the region of 30% of existing funding levels" in order to keep pace with demand, although "a substantial portion of that" could come from technology-driven productivity improvements in the agencies. Key problems still existing were maintaining consistent quality of data sets uniformly across data-supplying agencies and improved coordination among local, state, and federal government agencies. Three final recommendations covered greater adoption of standards for data transfer, including more attention to metadata, guaranteeing fair competition between data suppliers, and establishing performance measurement criteria to better gauge how efficiently resources are used, i.e., creating measurable success criteria.

Interestingly, the Price Waterhouse study began with a survey of 85 major data suppliers and 350 major users, to determine which methodology to use for the study itself, since the volume and quality of the data supplied would indicate which was best. The survey results led them to choose the cost-effectiveness approach as opposed to a traditional CBA, partly because detailed cost information was provided, but most benefit data were "of a qualitative nature." They then supplemented the analysis with seven case studies from different geographic areas and themes: law enforcement, public utilities, local council operations, health care, education, natural resources management, and mining. The study remains useful today.

6.5.2   *OXERA economic contribution of GI (1999)*

The report by Oxford Economic Research Associates Ltd., U.K. (OXERA), published in May 1999 for the Ordnance Survey GB, was commissioned to provide evidential support for the importance of the role that OSGB and geo-spatial data (topographic mapping, in this case), played in the economy of the U.K. as a whole. The economic value of OSGB as the primary map-producing agency in the U.K. was defined in the report as "the contribution which OS makes to the Great Britain economy as a producer of final and intermediate products and services, as a purchaser of intermediate products and services and ... as the provider of geographic information (GI) in the national inter-est." Tellingly, the study also begins (paragraph 2) with the warning "mon-etary values provided are ... broad indicators of the scale of the contribution of OS to Great Britain's economy. Given the lack of empirical evidence for a study of this kind, ... conclusions are reached on the basis of both qualitative and quantitative assessments." (So perhaps a good case for a cost-effective-ness analysis, as opposed to a traditional CBA?)

As to methodology, the study categorizes benefits as gains of three types: increases in efficiency, increases in effectiveness, and provision of new prod-ucts and services. These are achieved by reducing processing and search costs, reducing waste by better scheduling, reducing uncertainty for more efficient service delivery, and matching products and services to user needs. The report then assumes that "development of computer-based GIS ... has increased the efficiency and effectiveness with which GI is used throughout the economy." The study (p. 9) does acknowledge that there are many uses of GIS that generate significant benefits, not all of which are monetary, e.g., in health provision, social services, etc.

The OXERA study begins with the statement, page 1, that in 1996, when OS operating costs were approximately U.K.£78 million, its "products and ser-vices contributed to 12–20% of gross value added (GVA) in the UK, equal to UK£79 to UK£136 billion GVA." Even taking the lowest GVA figure, this indi-cates a raw cost:benefit ratio of 1,000:1. This calculation is further explained by the line: "this economic contribution of OS comes, in the main, through the use of OS products and services as a primary input into the production of several key sectors of the economy." Some in the GI industry, and those engaged in the policy and politics of SDIs, try to use the OXERA study as a proxy for a more traditional CBA proving that almost any level of investment in an SDI is warranted. After all, who could argue with a cost:benefit ratio of 1,000 to 1.

Sadly, this association between use of maps or other GI provided by OSGB and the value of the economy does not stand up to closer scrutiny, as it pre-sumes that this GVA is possible only because the maps or GI exist, and that there are no competing alternatives that could deliver the same functionality except by use of topographic GI from OS. More to the point, as more than one study has warned, as soon as cost:benefit figures become too good to be true, it is time to start questioning the methodology, statistics, or analysis used in their calculation. Most decision makers responsible for significant investments in projects the size of a national SDI simply do not believe such apparently wonderful ratios, as they are so far out of the ordinary range typically encountered that they seem immediately suspicious, even if they are factual.

6.5.3   PIRA European PSI exploitation report (2000)

The goal of the PIRA International Ltd. study (PIRA, 2000) was not to conduct a CBA for SDI creation, but rather to examine market size for public sector information (PSI) in Europe, compare this to global competitors, e.g., the U.S., in the information marketplace, and to make recommendations as to how Europe could better its position in that marketplace. The reason that we include the PIRA study in this report is to introduce the definition for value of information that was adopted by PIRA in conducting its study. Also, the study found that the value of the GI sector, at 38 billion euro, was the single largest sector for the projected European information market size for PSI of 68 billion euro, with the next nearest sector (economic and social data) reaching only 11.8 billion euro. By comparison, the value assigned to the U.S. information market was 750 billion euro in 1999, the discrepancy for two regions of approximately the same population being ascribed to the open exploitation policy for most PSI in the U.S.

   As to assigning value to information, PIRA's methodology differentiated between investment value and economic value. The former is what governments invest in acquiring PSI, while the latter represents the portion of national income "attributable to industries and activities built on the exploitation of PSI" (PIRA, 2000, p. 15), i.e., the value added by PSI to the economy as a whole. Economic value far surpassed investment value (an average figure of 68 billion euro compared to 9.5 billion euro for investment), but the traditional source for economic value figures (national accounts information of traditional industries) is not available for the information marketplace. Hence, the first assumptions already creep into the analysis in that "estimates of the value added by users to PSI … provide figures for the economic value of PSI." Whereas investment value (relating directly to costs spent in acquiring PSI) was quite accurately estimated at 9.5 billion euro (of which, again, GI was the largest single sector at 37%), the economic value figure used is actually a central estimate (not a simple average) based on a range of 28 billion to 134 billion euro. As with cost:benefit ratios in the stratosphere, one also begins to question entire economic analysis reports built around assumptions leading to such widely varying values for one of the key components of the analysis, i.e., economic value.

6.5.4   *INSPIRE extended impact assessment (2004)*

An early draft report from the INSPIRE Impact Analysis Working Group (undated) examined alternative methodologies to develop a process of systematic analysis of the likely impacts of the INSPIRE vision. Note that an impact analysis is supposed to identify positive and negative impacts of proposed policy actions and alternatives, of which a cost–benefit analysis is one tool. The working group selected a general equilibrium model (GEM) to assess the social and economic impacts of INSPIRE rather than other options (multimarket model, direct costs compliance model). The GEM attempts to define the economy as a whole (or selected aspects of it), then the implementation of INSPIRE by policy alternatives is assessed with regard to its effect on the economy (model), i.e., what is the new equilibrium state following application of the policy. The difference between the two states, if it can be allocated to the INSPIRE policies (data pricing, availability, quality assurance, etc.), is then used to calculate the net increase/decrease in welfare for each sector of the economy identified in the model.

This model approach was to be used to examine three questions:

1. Will those who gain most from implementation of INSPIRE be able to fully compensate those who lose — and still remain better off?
2. Who are the gainers and losers for different INSPIRE policy alternatives, and how much do they gain or lose by each alternative?
3. How does a particular sector of the economy fare under different INSPIRE policy implementations?

Question 1 is an efficiency measurement and the foundation of a traditional cost–benefit analysis that dictates that INSPIRE's net effect on society should be positive. Sadly, lack of statistical information needed for this analysis with any level of accuracy or credibility greatly hindered this approach. The two final questions relate to the distributional consequences of alternative INSPIRE implementations that can be examined using the distributional analysis approach. This approach, unlike the traditional CBA, examines the distribution of impacts of many INSPIRE measures. Both methods (CBA and distribution analysis) would be needed to best estimate the impact of alternative policies to implement INSPIRE across Europe — or for a national SDI as well.

A key aspect of applying the GEM is to have a baseline for how the world or economy looks without the planned policy measures, to be compared later to one that forecasts the impact of such measures. Differences to be measured include changes in employment levels, taxation generated, levels of commercial activity, etc. Unfortunately, much work is involved in defining this baseline, especially for all sectors that could be impacted by something as all-pervasive as geographic information (e.g., "80% to 85% of all PSI has a geographic element"). Get the baseline wrong and the rest of the analysis will be of little worth.

Equally unfortunately, when attempting to do a CBA for INSPIRE, the working group found that "a given INSPIRE measure or policy may produce many different benefits, but it is rarely possible to obtain a single, comprehensible value estimate for the collection of effects" (p. 19). Coupled with the lack of "detailed statistical input" even at the national level, let alone harmonized in a meaningful way for the whole of Europe (as INSPIRE is pan-European), one begins to see why many experts give little credence to the figures that were finally arrived at, then reworked — twice — in later versions of the report. One could be forgiven for concluding that the real goal seemed to many outsiders to be "get the costs low enough to not scare off national governments," as opposed to producing a valuable, evidence-based CBA.

The final impact assessment report for INSPIRE published in September 2003 does contain whole sections on investment costs and qualitative and quantitative benefits. In a concluding chapter, the report predicts annual costs across the whole EU (not attributed by country or region) of 200 to 300 million euro, compared to quantifiable benefits of 1.2 to 1.8 billion euro, a computed benefit:cost ratio of 4:1 (worst case) to 9:1 (best case), without making additional allowance for those qualitative benefits that are not included in the figures above. When EU member states found these cost figures still high, a further analysis in 2004, based on reduced scope for INSPIRE and some new assumptions on timing and coverage, resulted in: "The revision of the basic assumptions reduced costs from a range of 200–300 m€ down to a range of 125-183 m€. On top of that, the revision of the scope yields an additional reduction in costs from the range of 125–183 m€ down to a range of 93–138 m€" (Dufourmont, 2004, p. 11). The predicted benefits were reduced from the initial estimate of 1.2 to 1.8 billion euro to a new estimate of 0.77 to 1.15 billion euro, still achieving handsome benefit:cost ratios of from 5.6:1 (highest cost and lowest benefit) to 12.4:1 (lowest cost and highest benefit). At least none of these benefit:cost ratios seem wildly out of synch with ratios from national SDI CBA studies over the past decade.

6.5.5   U.S. national map cost–benefit analysis (2004)

A cost–benefit analysis for the U.S. The National Map, released in May 2004, provided an analysis spanning the 30 years planned for this ambitious program to bring the American topographic map base fully up to date over the next three decades (Halsing et al., 2004). Today, it is estimated that as much as 50% or more of the topographic data that are freely available from agencies such as USGS, under the U.S. Freedom of Information Act, are as much as 25 years out of date. The National Map is a project designed to correct this situation through budget increases for USGS and contributions of large-scale, up-to-date data from local, county, and state agencies over the next 30 years. Performing a sensitivity analysis incorporating more than 60 scenarios (50 runs each) indicates a net present value (NPV) of roughly U.S.$2 billion (with a standard deviation of U.S.$492 million, expressed in 2001 dollars). NPV does not turn positive until 14 years into the program, which would certainly have many politicians quaking across Europe if faced with similar figures for INSPIRE.

The methodology used for the CBA distinguishes costs and benefits of geospatial data from those of applications for the data, compares the state of the world with The National Map to one without it (see GEM discussion relating to INSPIRE impact analysis, above), assumes that uses of spatial data will increase over time (partly due to the very existence of a better-quality national map), and takes account of the varying ability of customers to make effective use of The National Map data. Predicted benefits are all those that have already been outlined elsewhere in this text.

Three main alternatives for creating The National Map were tested:[2]

- Create The National Map over 10 years with an incremental budget of $25 million per year (i.e., in addition to 2001 budget levels for the organizations involved).
- Create The National Map by diverting other USGS funds to this program, for which no guarantees are then made on when it would be complete, how complete, how accurate and consistent, etc.
- Do not create The National Map at all.

Note that these alternatives are not all that dissimilar to the five policy options explored in the INSPIRE extended impact assessment, which ranged from do nothing to voluntary cooperation of member states, two different levels of framework backed by an EU directive, or an EU regulation stating how member states will implement INSPIRE standards and infrastructure.

In performing the analysis for The National Map, the investigators first created a framework for the analysis by specifying their alternatives (three), enumerating their assumptions (eight), and proposing and explaining/justifying a specific economic model to be followed for estimating benefits (theories and formulae to arrive at net present benefit). With this analytical framework in place, they then developed a system and methodology to account for changes in the variable in their simulation mode, which was to run for 30-year periods.

Finally, regarding this analysis and the method of presenting its results, the reader is not faced with a single benefit:cost ratio on which an investment decision is to be made. Rather, a complete explanation is presented, scenarios are constructed and tested using multiple runs of a simulation program (fully specified in an annex to the report), and the financial results can be followed year on year, as variables change over time. Considering that the simulation (NB-Sim) was created with a sophisticated modeling package (which permits even fuzzy concepts to be analyzed) and accommodates more than 3,000 U.S. counties, arrayed across three tiers of sophistication, one feels more confident of the predicted outcomes.

6.5.6   *NASA/Booz Allen Hamilton: interoperability ROI (2005)*

The Booz Allen Hamilton (2005, p. 4) study of return on investment (ROI) for implementing geospatial interoperability technology ("the ability of two different software systems to interact with geospatial information") based on open standards is included in our research for the following reasons. First, because it is relatively recent (April 2005). Second, while not looking specifically at SDI from the national infrastructure point of view, it highlights the cost–benefit of implementing the sort of geospatial data interoperability that is one of the ultimate stated goals of most SDI visions, policies, and strategies, i.e., increasing ease of access to widely distributed GI resources.

The study uses a form of multicriteria analysis that combined financial cost analysis with value-based benefit assessments, assigned by experts and stakeholders to quantify the value of geospatial interoperability standards and to determine for whom, and when, different benefits accrue. The analysis then applies the methodology to different case studies that were investigated early on in the project. Eventually one case study for a project that used a high degree of open geospatial standards was selected, accompanied by a second that used few open standards.

The geographic VMM (GeoVMM) methodology used in the analysis is a version of the value measuring methodology (VMM) adapted for analyzing geospatial information projects, combining the cost–benefit, value judging, and risk analysis features of an advanced form of multicriteria analysis. GeoVMM was developed by the U.S. consulting firm Booz Allen Hamilton working with academics affiliated with Harvard University's Kennedy School of Government under contract to the U.S. government. The methodology assesses costs, benefits, and risks for five major government stakeholder groups: direct user, government financial, government operational and foundational, social and political, and strategic.

In the first stage of a four-stage process, an objective decision-making framework is created in which the cost, value, and risk structures are defined and agreed upon by stakeholders working with experts. This is critical since, if these three main structures are not accepted by the decision makers who receive the results of the analysis, then the results lose credibility, regardless of how much work goes into the later stages of the process.

Because VMM does not assign monetary figures to benefits, but rather looks at benefits in the structure of a range of agreed-upon values, it is important to get the value structure correct, i.e., agreed upon by both stakeholders and decision makers. The value structure is formed in two layers, the first of which comprises value factors important to the five categories of government stakeholders mentioned earlier. The value factors must be prioritized by the decision makers who will be most affected by the analysis, including key stakeholders and funders.

The second layer of the value structure comprises detailed subcategories that will appear in each main header, defined by project-level staff typically working with experts and in groups. For example, under a main value category of direct user value might appear subcategories of data availability, ease of use, and broad data sharing capabilities. Specification of these more detailed categories includes a metric, a target, and a normalized scale for making comparisons. The metric is needed in order to measure whether an initiative has delivered the expected benefits. Translating (or normalizing) performance measurements onto a single scale permits comparison of both objective and subjective measures of value. For the NASA ROI study, analysts worked with representatives of the user and partner communities in prioritizing the benefits within the specific second-layer value factors, assigning each with a weight and developing corresponding metrics.

In the second step of the analysis, the two case studies mentioned above were examined with respect to most likely costs, benefits, and risks based on the framework developed in step 1. In step 3, pulling together the information, the financial measures were calculated together with value, cost, and risk scores. Two decision metrics for each alternative were also produced: the return on investment (ROI) and an index reflecting the level of benefits, or value, achieved for each alternative. Alternatives were then compared to the best case. The value index is the value score of an alternative divided by the investment cost of that alternative, which avoids comparing apples and oranges. The comparison between alternatives is possible only because the two alternatives were analyzed in prior steps using the same formally specified decision framework against which all values were converted into a single (100%) scale. Note how dependent the methodology is on the completeness and quality of planning and analysis in the prior steps, especially on how well the decision framework was set up and agreed upon by stakeholders.

In step 4, the outputs of GeoVMM were used to communicate the value to stakeholders.

The study demonstrated to NASA the value of supporting the geospatial interoperability standards. Standards-based projects were shown to have a 119% ROI over the program that did not implement standards. One dollar invested in open standards-based projects nets $1.19 in savings in operations and maintenance compared to projects not based on open standards. This is called the savings-to-invest ratio. Standards were therefore found to lower transaction costs for sharing geospatial data when semantic agreement was reached between parties; e.g., the higher implementation costs for case study 1 (the standards-based project) are combined with considerably lower operations and maintenance (O&M) costs. This project also saved 26% of the overall cost compared to the project that did not adopt open standards. Stated differently, for every $4 spent on projects based on proprietary platforms, the same value could be achieved for $3 investment using open standards. Risk-adjusted transactions costs were also 30.3% lower for the open standards-based project.

The non-open standards project had a 50% higher risk in acquisition and implementation costs. All costs, from planning and development, to acquisition and implementation, to maintenance and operations, were significantly higher for the project using non-open standards, i.e., 27, 33, and 60%, respectively, than for the open standards-based project. One conclusion drawn by the researchers was that "use of proprietary models limits the flexibility and adaptability of the program over time (Booz Allen Hamilton, 2005, p. 5)."

However, the study also noted that the most successful project (case 1, using open interoperability standards) had the highest initial start-up costs. This was typically due to the level of extra training that was required to use the open standards, create quality metadata for the data resources in the project, and similar activities not related directly to creating or using the information system itself. The project also took longer to deliver because of this. Yet all these initial costs were more than recouped over the life of the project, which was also found to be much more adaptable to changing requirements as time progressed.

We offer the summary of this report as a prime example of how and why all organizations embarking on creating a new geospatial information system — and the SDI to underpin those applications — should certainly adopt open standards. Such standards are proliferating today thanks to the work of the Open Geospatial Consortium, Inc. (OGC) and its global affiliates, plus the work of ISO Technical Committee 211 (TC211), not to mention the past work of the U.S. FGDC, whose initial standards from 1998 were adopted widely around the globe (including in the U.K.), and current work of CEN TC287 in adopting European profiles of the ISO TC211 standards. The message is clear — use proprietary geospatial standards at your own risk, especially if looking to the future.

6.6   Conclusions and recommendations

In this chapter we have presented a number of topics dealing with spatial data infrastructure policy and strategy, and the cost–benefit analyses that are typically undertaken to justify the often considerable investments projected to implement an SDI. We close out the chapter with some simple recommendations based on firsthand experience in consulting with a range of regional (subnational), national, and transnational governments and government organizations over the past few years. These recommendations include, in priority order:

- Maintain openness and transparency throughout the SDI process when creating the SDI product, for both policy and strategy development.

- Involve as many stakeholders as practically possible from as early as possible in the exercise, and also in follow-on activities, including monitoring for success.
- Do not be afraid to set success criteria, even if these are not always quantifiable, as long as they are agreed upon with decision makers, funders, and major stakeholders.
- Perform information audits that will indicate what data are held, why, how they are used, and how often, with users' own estimates of the cost and benefit of having vs. not having that information readily available.
- Do not be afraid to request, even to demand, rigor from the experts conducting your cost–benefit analyses, and select an appropriate methodology suited to the information infrastructure environment in which you work.
- Use open standards wherever and whenever possible, as a central platform of your SDI visions, policies, and strategies.

Yet when all is said and done, more than one high-level government official in charge of SDI development at the national level has informed us that they expect further (or even initial) development of their SDI to be driven by a leap of faith, that it must be good for the society and the economy, regardless of what CBA studies, business cases, or any other analytical process tells them.

References

AGI Cymru. 2003. Geographical Information Strategy Action Plan for Wales. Association for Geographic Information, Wales Region, for the Welsh Assembly Government. http://www.agi.org.uk/SITE/UPLOAD/DOCUMENT/Reports/GIS_strategy_for_wales_english.pdf (accessed March 10, 2005).

ANZLIC. 2007. *Australian Spatial Data Infrastructure.* Australia and New Zealand Land Information Council, Canberra, Australia. http://www.anzlic.org.au/infrastructure_ASDI.html (accessed January 10, 2007).

Bawden, D. 1996. Information policy or knowledge policy? In *Understanding Information Policy*, I. Rowlands (Ed.). Bowker-Saur, London.

Blakemore, M. and R. Dutton. 2003, November. e-Government, e-Society and Jordan: Strategy, Theory, Practice, and Assessment. *First Monday*, 8. http://firstmonday.org/issues/issue8_11/blakemore/index.html (accessed November 4, 2003).

Booz Allen Hamilton. 2005, April. Geospatial Interoperability Return on Investment Study. National Aeronautics and Space Administration, Geospatial Interoperability Office. http://gio.gsfc.nasa.gov/docs/ROI%20Study.pdf (accessed December 18, 2005).

Bregt, A. and J. Crompvoets. 2005. Spatial data infrastructures: hype or hit? In *GSDI-8 Conference Proceedings*, April 2005. http://gsdidocs.org/gsdiconf/GSDI-8/papers/ts_36/ts36_01_bregt_crompvoets.pdf (accessed January 22, 2007).

Bridgman, P. and G. Davis. 2004. *The Australian Policy Handbook*, 3rd ed. Allen & Unwin, St. Leonards, NSW, Australia.

Burger, R. 1993. *Information Policy: A Framework for Evaluation and Policy Research.* Ablex Publishing Corporation, Northwood, NJ.

Burrough, P. and I. Masser. 1998. *European Geographic Information Infrastructures Opportunities and Pitfalls: GISDATA 5.* Taylor & Francis, London.

Cabinet Office. 2005, March 18. *e-Government Interoperability Framework*, version 6.1. U.K. Government, Cabinet Office, e-Government Unit, London. http://www.govtalk.gov.uk/documents/eGI%20v6_1(1).pdf (accessed January 22, 2007).

Cabinet Office. 2006, August 29. *e-Government Metadata Standard*, version 3.1. U.K. Government Cabinet Office, e-Government Unit, London. http://www.govtalk.gov.uk/documents/eGMS%20version%203_1.pdf (accessed January 22, 2007).

Clinton, W. 1994, April 13. Coordinating Geographic Data Acquisition and Access: The National Spatial Data Infrastructure, Executive Order 12906. *Federal Register*, 59, 17671–17674. http://frwebgate1.access.gpo.gov/cgi-bin/waisgate.cgi?WAISdocID=5671913174+0+0+0&WAISaction=retrieve (accessed October 18, 2000).

Copeland, T. and V. Antikarov. 2001. Real Options: A Practitioner's Guide, TEXERE, New York, NY.

Corle, F. 2004, June. STIA President Fred Corle Testifies before Congress. *GeoWorld*. http://www.geoplace.com (accessed July 16, 2005).

Craglia, M. 2003. Contribution to the Extended Impact Assessment of INSPIRE. Environment Agency, U.K. http://inspire.jrc.it/reports/fds_report.pdf (accessed January 20, 2007).

Crompvoets, J. 2006. *National Spatial Data Clearinghouses: Worldwide Development and Impact.* Wageningen University, Wageningen, NL.

Crompvoets, J. 2007. *Multi-View Frameworks to Assess NSDIs, workshop report.* Wageningen University, Wageningen, NL. http://www.grs.wur.nl/UK/Workshops/Multi-view+framework+NSDIs/Program/ (accessed June 5, 2007).

Dufourmont, H. 2004. *Extended Impact Assessment of INSPIRE Based on Revised Scope.* European Commission, DG Eurostat, Luxembourg. http://inspire.jrc.it/reports/inspire_extended_impact_assessment.pdf (accessed April 20, 2007).

EU. 2007, April 25. Directive 2007/2/EC of the European Parliament and of the Council of 14 March 2007 Establishing an Infrastructure for Spatial Information in the European Community (INSPIRE). *Official Journal of the European Union* (Luxembourg). http://www.ec-gis.org/inspire/directive/l_10820070425en00010014.pdf (accessed April 30, 2007).

European Commission. 1998, August. Geographic Information in Europe: A Discussion Document. European Commission, DG XIII/E (now DG Information Society, Information Market Directorate). 1998. http://www.ec-gis.org/copygi2000/gi2000/discussion98_.html (accessed April 22, 2007).

European Commission. 2000. *Directive 2000/60/EC of the European Parliament and of the Council of 23 October 2000 Establishing a Framework for Community Action in the Field of Water Policy.* Office of Official Publications of the European Communities, Luxembourg.

European Commission. 2001, December 20. *ESDI Organisation and E-ESDI Action Plan*, version 0.13, final. Directorate General Environment, Directorate B — Territorial Dimension, European Commission, Brussels. http://www.ec-gis.org/inspire/reports/ESDI_Action_plan1aa13.pdf (accessed February 20, 2006).

European Commission. 2002a, April 11. *Memorandum of Understanding between Commissioners Wallström, Solbes, Busquin: Infrastructure for Spatial Information in Europe (INSPIRE).* European Commission, Brussels. http://www.ec-gis.org/inspire/reports/MoU.pdf (accessed February 10, 2006).

European Commission. 2002b. *Directive 2002/207/EC on the Re-Use and Commercial Exploitation of Public Sector Documents.* Office of Official Publications of the European Communities, Luxembourg.

FGDC. 2004, June 15. *Towards a National Geospatial Strategy and Implementation Plan.* NSDI Future Directions Planning Team, FGDC, Washington, DC. http://www.fgdc.gov/policyandplanning/future-directions/reports/FD_Final_Report.pdf (accessed January 4, 2006).

FGDC. 2007a. *NSDI Framework.* FGDC, Washington, DC. http://www.fgdc.gov/framework/ (accessed May 12, 2007).

FGDC. 2007b. *Strategic and Business Plan Development in Support of the NSDI Future Directions: 50 States Initiative.* FGDC, Washington, DC. http://www.fgdc.gov/policyandplanning/50states/50states (accessed May 14, 2007).

GI Panel. 2007. GI Strategy for the United Kingdom. Geographic Information Panel, U.K. government. http://www.gipanel.org.uk/gipanel/gistrategy/index.html (accessed May 31, 2007).

Groat, C. 2003. The National Map: a continuing, critical need for the nation. *Photogrammetric Engineering and Remote Sensing Journal,* 69. http://www.asprs.org/publications/pers/2003journal/october/highlight.pdf (accessed May 25, 2007).

Groot, R. and J. McLaughlin. 2000. *Geospatial Data Infrastructure Concepts, Cases and Good Practice.* Oxford University Press, Oxford.

Halsing, D., K. Theisen, and R. Bernknopf. 2004. *A Cost-Benefit Analysis of The National Map,* Circular 1271. U.S. Department of the Interior, U.S. Geological Survey, Reston, VA. http://nationalmap.gov/nmnews.html (accessed January 4, 2006).

ILRI. 1995. *Livestock Policy Analysis,* ILRI Training Manual 2. International Livestock Research Institute, Nairobi, Kenya. http://www.ilri.org/html/trainingMat/policy_X5547e/x5547e00.htm#Contents (accessed, May 10, 2007).

Kelman, S. 1987. *Making Public Policy: A Hopeful View of American Government.* Basic Books, New York.

Kelmelis, J., M. DeMulder, C. Ogrosky, N. Van Driel, and B.J. Ryan. 2003. The National Map: from geography to mapping and back again. *Photogrammetric Engineering and Remote Sensing Journal,* 69. http://nationalmap.gov/report/PERS_article_forprint.pdf (accessed May 25, 2007).

Koontz, L. 2004. *Geospatial Information: Better Coordination and Oversight Could Help Reduce Duplicative Investments.* Testimony before the Subcommittee on Technology, Information Policy, Intergovernmental Relations and the Census, House Committee on Government Reform, U.S. General Accounting Office, Washington, DC. http://www.gao.gov/new.items/d04824t.pdf (accessed June 6, 2007).

Longhorn, R. 2002. A comparison of spatial information access policies of transnational environmental modeling and global climate change programs. In *Geoinformation for European-Wide Integration,* T. Benes (Ed.). Millpress, Rotterdam, NL, pp. 305–313.

Longhorn, R. 2006, November 29. Summary of comments from GITA board members regarding NSDI, unpublished e-mail exchange. Available on request from the author at ral@alum.mit.edu.

Lowi, T. 1972. Four systems of policy, politics and choice. *Public Administration Review,* 32: 298–310.

Lukas, V. 2004. Final Report: The National Map Partnership Project. The National Map Partnership Project Core Team, USGS, U.S. Department of the Interior. http://nationalmap.gov/report/NSGIC_TNM_Report_102705_V6.doc (accessed (July 28, 2006).

Masser, I. 2005. *GIS Worlds: Creating Spatial Data Infrastructures.* ESRI Press, Redlands, CA.

Masser, I. 2007. *Building European Spatial Data Infrastructures.* ESRI Press, Redlands, CA.

Montgomery County Council, Maryland. 1999, April 12. *Geographic Information System Cost/Benefit Assessment Report M-NCPPC*. Montgomery County Council Management and Fiscal Policy Committee.

NCLIS. 2001, January 26. *A Comprehensive Assessment of Public Information Dissemination*, final report, executive summary. U.S. National Commission on Libraries and Information Science, Washington, DC. http://www.netcaucus.org/books/egov2001/pdf/assessme.pdf (accessed June 2, 007).

Nebert, D. 2000, July 6. *Developing Spatial Data Infrastructures: The SDI Cookbook*, version 1.0. GSDI Association. http://www.gsdi.org/pubs/cookbook/chapter01.html (accessed January 10, 2006).

NGDF Management Board. 1999, December 17. *National Geospatial Data Framework: The UK Standard Geographic Base*, summary and status report, version 1.1. Ordnance Survey GB, Southampton, U.K.

NRC MSC. 1990. *Spatial Data Needs: The Future of the National Mapping Program*. Mapping Science Committee, National Research Council, National Academy Press, Washington, DC.

NRC MSC. 1994. *Promoting the National Spatial Data Infrastructure through Partnerships*. Mapping Science Committee, National Research Council, National Academy Press, Washington, DC.

ODPM. 2005. *Consultation on the Future of the National Interest Mapping Services Agreement (NIMSA)*. ODPM, London. http://www.propertylicense.gov.uk/embedded_object.asp?id=1144577 (accessed July 7, 2005).

Ogrosky, C.E. 2003. The National Map foreword: national mapping examined: an introduction. *Photogrammetric Engineering and Remote Sensing Journal*, 69. http://www.asprs.org/publications/pers/2003journal/october/foreword.html (accessed May 25, 2007).

OMB. 2002. *Coordination of Geographic Information and Related Spatial Data Activities*, Circular A-16, revised. OMB, Washington, DC. http://www.whitehouse.gov/OMB/circulars/a016/a016_rev.html (accessed May 25, 2007).

Onsrud, H. 2007. *Research and Theory in Advancing Spatial Data Infrastructure Concepts*. ESRI Press, Redlands, CA.

Ordnance Survey. 2004, September. The Digital National Framework, a white paper by the Ordnance Survey. http://www.dnf.org/Papers/DNFWhitePaper.pdf (accessed September 15, 2007).

Orna, E. 1999. *Practical Information Policies*. Gower Publishing Ltd., Aldershot, U.K.

Osborne, M. and J. Pepper. 2006. Extending the UK's digital national framework offshore. In *Proceedings of the CoastGIS 2006 Conference*. http://www.coastgis.org/past_conferences.htm (accessed May 25, 2007).

OSNI. 2002. *A Geographic Information Strategy for Northern Ireland*. GI Strategy Project Team, Ordnance Survey Northern Ireland, Belfast. http://www.mosaic-ni.gov.uk/contentDocs/Links%20and%20Resources/pdf/GI%20Strategy.pdf (accessed January 15, 2005).

OXERA. 1999. The Economic Contribution of Ordnance Survey GB. OXERA Ltd., U.K. http://www.ordnancesurvey.co.uk/aboutus/reports/oxera/oxera.pdf (accessed July 10, 2006).

PIRA. 2000. Commercial Exploitation of Europe's Public Sector Information, executive summary and final report. PIRA International Ltd and University of East Anglia for the European Commission, DG Information Society. http://ec.europa.eu/idabc/en/document/3538/5662 (accessed July 10, 2006).

Price Waterhouse. 1995. Australian Land and Geographic Data Infrastructure Benefits Study. Price Waterhouse Economic Studies and Strategies Unit, Australia New Zealand Land Information Council (ANZLIC). http://www.anzlic.org.au/get/2358011751.pdf (accessed September 18, 2007).

Rajabifard, A. 2002. Diffusion of Asia-Pacific spatial data infrastructure: from concept to reality. In *Proceedings of Seminar on Implementation Models of Asia and the Pacific Spatial Data Infrastructure (APSDI) Clearinghouse*, Negara Brunei, Darussalam, April 17, 2002. http://www.sbsm.gov.cn/pcgiap/brunei/seminar/apsdi.pdf (accessed June 6, 2007).

Rogers, E.M. 1995. *Diffusion of Innovations*, 4th ed. Free Press, Simon & Schuster International, New York.

Rowlands, I. (Ed.). 1996. *Understanding Information Policy*. Bowker-Saur, London.

Salisbury, R. 1968. The analysis of public policy: a search for theories and roles. In *Political Science and Public Policy*, A. Ranney (Ed.). Markham Publishing, Chicago, pp. 151–175.

Scottish Executive. 2004. Model Policies in Land Use Planning in Scotland: A Scoping Study. Scottish Executive, Ministry for Finance and Public Service Reform. http://www.scotland.gov.uk/Publications/2004/06/19512/39141 (accessed May 28, 2007).

Scottish Executive. 2005. One Scotland — One Geography: A Geographic Information Strategy for Scotland. http://www.scotland.gov.uk/Resource/Doc/57346/0016921.pdf. (accessed February 10, 2006).

SDI Africa. 2004. SDI Africa: An Implementation Guide. UN Economic Commission for Africa, GSDI Association, EIS-Africa. http://geoinfo.uneca.org/sdiafrica/Chap_HTML/00Preamble.htm (accessed May 22, 2007).

Tolbert, C. 2002. Rethinking Lowi's constituent policy: governance policy and direct democracy.*Environment and Planning C: Government and Policy*, 20: 75–93.

UNECA. 2001. *The* Future Orientation of Geoinformation Activities in Africa, position paper. Geoinformation Unit, Development Information Services Division (DISD), United Nations Economic Commission for Africa, Addis Ababa. http://geoinfo.uneca.org/sdiafrica/Reference/Ref0/Orientation_GI_Africa.pdf (accessed May 28, 2007).

USGS. 2007. *The National Map*. USGS, Reston, VA. http://nationalmap.gov/ (accessed December 18, 2005).

Van Loenen, B. 2006. *Developing Geographic Information Infrastructures: The Role of Information Policies*. DUP Science, Delft, NL.

Van Loenen, B. and B. Kok (Eds.). 2004. *Spatial Information Infrastructure and Policy Development in Europe and the United States*. DUP Science, Delft, NL.

Weiss, P.N. and P. Backlund. 1997. International information policy in conflict: open and unrestricted access versus government commercialization. In *Borders in Cyberspace*, B. Kahin and C. Nesson (Eds.), MIT Press, Cambridge, MA, pp. 300–321.

Wikipedia. 2007. http://en.wikipedia.org/wiki/Policy.

chapter seven

Conclusions and prospects

7.1   The debate is not concluded

We hesitate to use the term *conclusions* for this chapter. The fluidity of the information landscape is such that events continually challenge many of our beliefs and practices. However, there are observations and conceptual summaries that help to explain where we have come from, why, and hopefully offer some insight into where we will be going.

First, let us be quite clear — we are not biased one way or the other toward free or priced information. We straddle the fence on the fee or free debate until more research has been concluded, and not only via formal (objective) information econometrics or prejudice-laden (subjective) case studies or anecdotes, pro or con. The case for free information can be made on the basis of freedom of information principles, for the public good and delivering public value. Yet the very sector that conducts much of the research into information access and pricing, and writes about the results, namely, the higher education sector, has to date been one of the most restrictive information producers with regard to intellectual property rights (IPR), preferring to publish in expensive academic journals rather than freely on the Web. As Michael Geist argues, "The model certainly proved lucrative for large publishers, yet resulted in the public paying twice for research that it was frequently unable to access" (Geist, 2007).

There have been renewed calls globally for wider public access to research through an information commons. For example, the European Commission is allocating significant funding to the creation of open-access research output, setting aside 75 million euro to fund infrastructure and preservation of scientific information resulting from its Seventh Research and Technology Development (RTD) Framework Program on the principle that "access to research outputs should be accessible to all through open repositories after an embargo period" (JISC, 2007).

There are clearly some governments where a strategic decision has been taken to release data for the wider public good, as was the case for Canada in April 2007.* Geoconnections Canada announced that "the department's new no-fee policy will help the natural resources sector and others develop knowledge, introduce innovations, and improve productivity — giving Canadians the advantage to succeed." Similarly, the 2005 law and 2006

---

*   http://www.nrcan-rncan.gc.ca/media/newsreleases/2007/200728_e.htm.

decree (Government of Catalunya, 2005, 2006) governing use of cartographic and geographic information (GI) within the Spanish province of Catalunya establish basic principles for free access and use of geographic information created by regional government bodies and recorded in the official cartographic register of Catalunya.

These are brave attempts at stimulating the geospatial market, and success will be dependent upon two major issues: a sustainable funding stream and the ability to match data provision to market needs. In the fine print of the Geoconnections announcement there are important qualifications, i.e., "the new no-fee access policy applies to data that is solely owned by NRCan"* (Natural Resources Canada), and the Geogratis Web Portal,** through which free data are accessed, does point back to the Geoconnections*** portal where chargeable and nonchargeable data can be discovered. Similarly in Catalunya, geographic information at useful scales is made available for commercial use for a fee,**** defined as use of "cartographic data and cartographical information in all kinds of publications having a sale price to the public produced on paper ... on digital support or by telematic means" (ICC, 2007).

Within the context of the arguments we made earlier in this book, both the Canadian and Catalunyan initiatives can be interpreted as brave decisions to free up important GI in a way that can stimulate usage in both government and society generally and generate public good. However, it is clear that no assumptions are made that the public good will provide practical support for the tasks of data maintenance and enhancement that would be of benefit to the original data holders or future users. From the practical point of view, the GI authorities in Catalunya are already considering — with some trepidation — just how they will go about assuring the quality, consistency, and harmonization of data that are submitted to their official register from sources outside the direct control — and expertise — of the cartographic agency itself. Yet this form of feedback and official imprimatur is what their recently enacted and liberally-minded cartographic law specifically permits.

The public good that is indirectly generated by wider data use is an additional benefit resulting from the investments that are needed to maintain the free-of-charge initiative. It is without doubt that such financial support will involve sensitive and difficult negotiations should there be a spending squeeze in the future. At the time of the Canadian announcement (April 2007), the Canadian economy***** was showing strong growth, and these are just the conditions needed for governments to make a leap of faith into medium-term public subsidies. In the Catalunya case, the new law on cartographic information is only now being implemented, and funding streams

---

* http://www.nrcan-rncan.gc.ca/media/newsreleases/2007/200728_e.htm.
** http://www.geogratis.cgdi.gc.ca/geogratis/en/index.html.
*** http://geodiscover.cgdi.ca/gdp/index.jsp?language=en.
**** http://www.icc.es/web/content/en/common/icc/condicions_us_ciu.html.
*****http://www.fin.gc.ca/ECONBR/ecbr07-04e.html.

to support free access must be secured via an annual budgeting process from the regional government. Securing the level of funding needed is a continual battle for most tax-voted agencies, wherever located and regardless of the sector of government in which they operate, especially as users tend to want ever more in the way of products and services at ever lower costs, or even for free, to be achieved within fixed annual budget limits.

However, there is often a scale issue present in many free-of-charge spatial data infrastructures (SDIs) or for the type of data that is made freely available, even where charging regimes exist. For example, the European Union's regional SDI, embodied in the INSPIRE directive, focuses on data at a scale of 1:250,000 — not a scale known for its relevance to planning, vehicle navigation, or the utilities. Several global GI resources are readily available, many without restrictions on reuse, but at scales of 1:1 million or smaller (up to 1:5 million). Yet regional (subnational) and especially local authorities require and work with much more detailed data, typically at scales of 1:1,000 or 1:5,000 up to 1:25,000, for which they are often data owners, legal custodians, or major stakeholders. For example, the government of Valencia in Spain provides the gvSIG* portal, where open-source software is provided along with links to free data.** Yet even this facility does not counter the arguments we have made for fee or free. It shows how it is more possible to undertake free-of-charge initiatives where those funding free access are also data providers, application stakeholders, and, more importantly, direct beneficiaries. In that context, the indirect public benefit does have an identifiable cost–benefit to the funding organization.

The current fee or free contest is not unique to the GI sector. In other infrastructures, there is a move away from provision via subsidy to pay-for-use, especially where the subsidy has proved to be inadequate to meet the demand that arose within the free-access regime. This is happening, for example, with driving on public highways, such as congestion charging in cities (Millward, 2007) and wider proposals in the U.K. for real-time road use charging linked to GPS monitoring. These forms of paying for infrastructure are very unpopular with citizens, as evidenced by the 1.8 million U.K. road users who signed a petition decrying the proposal for real-time charging,*** but are very attractive for politicians, since they relink use with payment (Kablenet, 2007). Such moves can then be further linked to the downstream consequences of driving, for example, through carbon taxes that help to mitigate environmental damage. Paradoxically, while citizens are highly resistant to paying for driving directly, there is strong support for taxes on pollution by businesses (Bortin, 2007). Perhaps rather naïvely, the survey respondents do not realize that the taxes on business inevitably will be factored into prices, so they will pay the taxes indirectly anyway. Even in

---

*   http://www.gvsig.gva.es/index.php?id=que-es-gvsig&L=2.

**   http://www.gvsig.gva.es/index.php?id=mapas-libres&L=2.

*** http://www.timesonline.co.uk/tol/news/politics/article1459230.ece.

the U.S., home of many information market myths regarding free government information, the national road infrastructure includes both free and toll roads. The telephone infrastructure for decades incorporated free local phone calls to all, but the real costs were subsidized by long-distance phone call charges, whether you or someone else made those calls. Remember that "Ma Bell," the national Bell Telephone Company *de facto* monopoly, was not a charity or a not-for-profit corporation.

Countering some of the move to direct payment for specific use, there are bundling pricing options linked to the rapid convergence of communications devices and channels. Google has moved into telecoms and software that will compete with Microsoft's domination of business software (Helft, 2007). Even in the health industry infrastructure, which is probably much dearer to most readers' hearts than geographic information provision, multiple business models already exist globally and even within single nations. For example, a patient may receive free treatment for some medical conditions but not others, or be required to pay for some services or medicines and not others, or pay different prices depending upon how much medicine is needed and over what period of time, or whether an operation is performed next week or in 6 months. The point is that the geographic information market, even the generic information market, is not unique in being required to accommodate different value chains, pricing and charging regimes, or paradigms. Nothing is ever truly free — someone always pays — the emotionally charged debate is, of course, over who does the paying.

In Chapter 2, we looked at how difficult it is to attach any single value to geographic information, which itself has many definitions, as discussed in Chapter 1. In Chapters 3 and 4 we looked in depth at why and how information is priced, sometimes with little relationship to actual vs. perceived value, or exchange vs. use value. We acknowledged the often religious zeal surrounding rights of access to information. In Chapters 5 and 6, we acknowledged very pertinent arguments for making information available as widely as possible, looking at the different cost–benefit issues and methodologies that provide both qualitative and quantitative underpinning to arguments of faith about access to information. We have no problem with the broad arguments that say more information, used by more citizens, is good for society, even if we do not support a direct, *de facto*, linear relationship between the notions of more and benefits.

In the end, however, we argue that the crucial debate is not about price or charging regimes per se; it is about consistent resources for reinvestment and maintenance of information that is fit for a wide range of purposes, while at the same time maximizing the ability of information providers to respond to the widest possible constituency or market. This is a key point — perhaps the one message above all others that we would like readers to take away from this book. It underpins the background theme that runs through the book: there is no such thing as a free lunch. Rather, the real question is who pays for that lunch, when, how, and who benefits. We summarize the rationale for our views with cases that we studied in the first few months of 2007, a short period during which the volatility of events in information space was apparent, starting with Google Earth.

7.2   Google: a free lunch?

Google Earth is wonderful. It is free to use, but looking at it in February/ March 2007, is it really something that will overturn the status quo of mapping agencies and their overall dominance of the GI production market? We have already shown that even without Google Earth, the availability of good-quality official mapping information in Egypt was so poor that key actors in the market in effect declared independence and started to collect their own information. Google Earth presents challenges to official data suppliers within national borders who may not be up to the mark, while transcending borders by offering global access to information that may be censored in one state, for example, on secrecy or homeland security grounds, but available to any enemies who have access to the Internet. In stating that Google Earth challenges official GI providers who may not be performing their functions well today, perhaps we should qualify the timescale. While much of Google Earth's geographic information is image based, not current, and of unknown provenance, as an organization Google has created the infrastructure to deliver higher-quality GI as soon as it becomes economically feasible — and commercially sensible — to do so. Operating within an aggressive online business model across a range of services, not just for geographic information, Google could be a threat to underperforming mapping agencies for at least some portion of those agencies' lines of business, including for current clients within other government agencies.

In its operation, Google Earth follows a classic business pricing model; only it does it on a huge spatial scale. The licensing options* are clearly stated. Free data and free software are available on the portal. Then there are value-adding options available at prices increasing in orders of magnitude. For $20 (April 2007 prices) there is the Plus option offering facilities such as "Plug in your GPS device to see your current position in real-time, or import data from your trek to relive the adventure." For $400 the Professional tool offers a wide set of functions and value-adding facilities for a business. The Enterprise option offers enterprise-wide and market development solutions, and the price of the license is negotiated according to the business proposition — in effect a value-adding reseller and franchise process. Therefore, there is little in Google Earth that radically disturbs the existing pricing strategies for data. To date, Google Earth has not been a producer of original data, but is an intermediary reseller, having developed licenses with GI producers. Therefore, when we access the "free" Google Earth facility, our particular free lunch is paid indirectly by Google through other activities — and by

---

*   http://earth.google.com/products.html.

other users — via click-pay advertising and sales of nonfree versions of the software to higher-end users.

Google Earth also showed itself to be understanding of what could be termed its global corporate social responsibility. Faced with concerns from governments that sensitive information was being made too freely available, Google removed details of U.K. Army bases in Iraq (Harding, 2007) at the same time that its freely available information was being used by citizens in Iraq to identify, and navigate around, dangerous areas (Hussein, 2007). Thus, Google is being both socially responsible by providing the free resources, and politically responsible by not threatening the sovereign rights of a government. For example, access to the freely available Google Maps API* (application programming interface), enabling programmers to embed Google maps in their own Web pages with JavaScript, is introducing a new cohort of trained GI specialists in Iraq, and should the economy stabilize, some of them will develop commercial applications and enter into licenses with Google and local or national data suppliers.

There are two reasons for Google not to upset sovereign governments. First, a government could make things very difficult for Google to operate its various businesses within national borders, and not just Google Earth, but all Google desktop-type applications available today, all of which are available in both free and pay-to-use versions. Second, Google seems to accept the political dilemma faced by a government in which it is easier for a government to seem silly for removing information that could be found elsewhere on the Internet than to damage its image by leaving the information officially accessible and then being blamed for any resulting terrorist outrage. This is just one example illustrating that the politics of information provision are much more delicately contested than the pricing of information.

7.3   Other fee-or-free contests and challenges

In the early months of 2007, the contest between production and consumption of information continued to show uncertainty, business innovation, and political shifting. The battle over whether free is less accurate or trustworthy than fee continued to swirl around Wikipedia, with concerns that a key contributor had faked academic qualifications (Cohen, 2007a). There is, of course, no causal link between a free resource and faked qualifications, since there are similar problems in the paid information arena, as evidenced by numerous recent cases of highly respected — and very expensive — peer-reviewed scientific publications having to withdraw articles for which the underlying evidence was later proven to have been faked (Agence France-Presse, 2007; Marshall, 2000). Concerns over the accuracy of free resources such as Wikipedia led one U.S. educational institution to forbid students to use it in their studies (Cohen, 2007b). Such a policy seems to deny the

---

* http://www.google.com/apis/maps/.

contributions from students and staff in developing learning strategies that provide skills for information evaluation. It further seems to assume that all traditional approved sources are accurate, something that Alan Sokal and Jean Bricmont (1999) exploited when they produced *Intellectual Impostures* — a contrived parody of academic posturing that was received and approved as valid through the peer review process. If you are in an institution that does not forbid you to access Wikipedia, you can read more about their book and the outcomes on that free online resource.*

Another contest, the subconscious exchange of personal data for free resources, through online advertising, is exemplified by people who feel that the recipients of advertising should become organized. This involves people gathering information about their own Internet use, preferences, and characteristics using software that plugs in to their browser, and the "resulting profile can then be deposited in an online vault, where interested parties can pay to see it" (Economist, 2007). This sounds rather perverse, since we consciously auction our own information to people who provide us free resources, but in reality it is just another example of resource exchange.

In March 2007, Yahoo announced that it would abolish the 1-gigabyte limitation on e-mail storage, allowing now unlimited e-mail storage. When first introduced, Yahoo e-mail limited users to 4 megabytes. David Filo, cofounder of Yahoo, was quoted as saying, "People should think about e-mail as something where they are archiving their lives" (Reuters, 2007). The business strategy behind this is simple — use increasingly cheap storage to offer a carrot for users to remain with your service. Encourage them to store masses of data, and then provide them with new facilities to organize, process, communicate, and visualize the resulting information overload. The growth of Internet radio, such as the Pandora** service, a classic Internet model that provides a free service via online advertising that you accept, was apparently threatened by a U.S. copyright ruling that may double the copyright fee paid for each track of music (Cellan-Jones, 2007). This again confirms our argument that product or service providers who rely on indirect income streams, in this case online advertising revenue, face risks, especially when there are external regulatory uncertainties such as copyright fee rulings.

Finally, we return to one of the challenges identified in earlier chapters: Is it right for us to receive benefit from the free data in San Fransisco? This challenge concerns the development of the commons concept, whether it is for information or for software. At what stage do the providers of information remove their participation because others are profiting from it? Informed self-interest seems to have underpinned the development of Wikipedia, with the occasional presence of motives such as the five minutes of fame and attention seeking in the form of deliberate injection of errors into entries to get a rise out of a global audience. In the arena of open-source software,

---

*   http://en.wikipedia.org/wiki/Fashionable_Nonsense.
**   http://pandora.com/.

there have always been businesses that "profit from this volunteerism — but only if they don't get too greedy" (Fox, 2007). This situation resembles the conventional supply chain challenge for any business; i.e., annoy your suppliers enough, in this case volunteer programmers or free data providers, and you may lose some of them, which may then threaten your business viability. The challenge for any provider of information products or services based predominantly on access to free data is to plan for the risk of losing that access.

7.4   Final lessons

In the end there are some prevailing characteristics of the information markets that impinge on the globalization of geographic information production and consumption.

First, there is a growing mismatch between organization speed and market speed. In organization speed, we include the speed with which legislation and regulation activities of government react to or lag behind events, as well as their organizational ability to actually enact legislation and regulation, and build information resources that are relevant to the wider market. Driven, or enabled, partly by the speed of innovation in the media technology industries, information market speed will always exceed that of the ability of organizations and institutions to catch up with the latest innovation. In the cyber age, new information products and services are brought to market in a matter of months, while legislation and regulation are reactive and typically take years to implement.

Second, the importance and role that public sector information (PSI) plays in the economy will continue to be strong through its role in allocation of government resources and the measurement of government performance. This has been particularly evident in the measurement of e-government. In the geographic information arena, significant volumes of the GI used by public authorities are collected, updated, or maintained by commercial data providers, even though ownership may rest with the public body, and the trend globally is for ever more PSI to be collected by commercial actors. Agreeing on the intellectual property rights of such data is of paramount importance for both public bodies and their commercial contractors.

Third, national-level PSI will continue to be contested concerning its relevance and quality in relation to local-level needs. The ability of local organizations to collect high-quality geospatial data has never been greater, thanks to the availability of low-cost, high-resolution data-gathering technologies. The fact that it is then difficult to integrate a bricolage of local information resources into a coherent national one is not an issue for the local user, although integrating multiple local resources for local use remains an issue. The real issue for national agencies is that local-level data show national-level data to be in error and out of date, leading to projects such as The National Map (TNM) in the U.S. (Kelmelis et al., 2003). Via TNM, the

U.S. Geological Survey is attempting to update national GI coverage that is in some places more than 50 years out of date (Brown, 2002) by encouraging local government GI holders to contribute their current and large-scale data to the national database.

Fourth, a growing threat exists wherein PSI continues to be collected by government, directly or by subcontract, but where the only users of the data are organizations that are mandated to use the official data through an official process monopoly. As we saw with Egypt, the private sector has shown that it cannot and will not wait for government to produce official GI and has collected high-quality information itself. A similar process is happening in India, for example, with companies collecting city-level street and property information* because it is not yet available from government data producers. Since much government-level GI is already collected by third-party commercial firms, in both developed and developing nations, what is to stop other potential users of geospatial PSI from simply employing the same data collectors, operating to the same standards? This creates a situation that of course contravenes one of the underlying principles expressed in virtually every spatial data infrastructure vision and strategy, whether at the national, regional, or global level: do not duplicate data collection or "collect once, use often."

Fifth, there will be continuing challenges to the information and knowledge commons through the uncertain exercising of monopoly patents on a global scale. This is particularly true where patents start to gain control over ideas, business methods, and algorithms, as in some national jurisdictions today, but not others, and not just over physical devices or physical processes, the "inventions" originally envisioned in the Paris Convention for the Protection of Industrial Property in 1883, since then much amended. Yet as we moved from an agricultural to an industrial society, in which the Paris Convention made sense (and actually refers to industrial and agricultural processes in Article 1), to an information and knowledge society, such contentions were bound to multiply, not lessen, and have major impacts on how we access data in the future, and who can have access and under what conditions. Information and knowledge are the industrial raw materials of the information and knowledge societies and economies.

Sixth, the process of making geographic information available will engender ever more flexible strategies in the future. As with the provision of non-geospatial information, such as newspapers, some providers will try free, others fee, and yet more will try hybrid strategies wherein some form of partial free access locks customers into a service so that they are willing then to pay for other information and services, i.e., the Google approach, whether for Google Earth, Maps, Writely, or Spreadsheets. For government PSI producers, the real price and charging challenges will continue to be those of balancing often short-term (annual) government policy-based funding decisions, hardly conducive to long-term planning, with the real needs of information

---

*   http://www.biondsoftware.com/.

users in government itself, which are long term. That is before taking into consideration the myriad potential users outside government, who could use added-value geospatial PSI, if available from commercial providers who are far better equipped — and motivated — to produce such products than are government data holders.

In conclusion, as we said at the beginning of this chapter, there may be no conclusion. Rather, it is our heartfelt wish that readers of this book, whether from the government or industry, private citizens, or map hackers of the world, in developed or developing nations, join or reenter the various global debates on the issues raised in the preceding chapters with an open mind. While researching this book, we found that many of you hold very strong beliefs, even lifelong convictions, on several of these issues — value of information, access for free or fee, charging and cost recovery by government agencies. Yet the information market is one of the most rapidly changing market places in the world today, challenged perhaps only by the speed of innovation we see in the financial marketplace. The information market underpins the global information and knowledge societies — and their emerging economies — just as transport and electricity and early telecommunications infrastructure underpinned the agricultural and industrial societies and economies.

Remember that the Internet is less than 25 years old, and the Web barely 15, if one counts from Tim Berners-Lee and Robert Cailliau's Hypertext project at CERN in 1990 as the birth of Web technology. The way we create, access, merge, converge, electronically cut and paste, plagiarize, transmit, disseminate, use, and abuse information today was unthinkable even a decade ago — and this includes text, images, sound, video, music, and even online sign language for the deaf. If recent history is any guide, many paths will be followed in the future for information provision in ways and for uses, both divergent and convergent, that we can barely imagine today. So perhaps it is useful to keep an open mind on how all this information and these exciting new allied products and services are going to be paid for — and by whom.

There is no such thing as a free lunch. Yet that does not mean that you need to pay for all your own lunches — as long as you accept that someone is paying — and are willing to risk that your benefactor's funding stream does not disappear before that next lunch.

References

Agence France-Presse. 2007. Swedish Scientific Breakthrough was Faked. SeedMagazine.com. http://www.seedmagazine.com/news/2007/04/swedish_scientific_breakthroug.php (accessed April 19, 2007).

Bortin, M. 2007, February 23. Poll Finds Strong Support in Europe and U.S. for Polluter Taxes. *International Herald Tribune.* http://www.iht.com/articles/2007/02/22/news/poll.php (accessed February 26, 2007).

Brown, K. 2002. Mapping the future. *Science,* 298: 1874–1875.

Cellan-Jones, R. 2007, March 8. Royalties Threaten Internet Radio. BBC. http://news. bbc.co.uk/1/hi/technology/6430489.stm (accessed March 8, 2007).

Cohen, N. 2007a, March 12. After False Claims, Wikipedia to Check Degrees. *International Herald Tribune.* http://www.iht.com/articles/2007/03/12/business/wiki. ph. (accessed March 13, 2007).

Cohen, N. 2007b, February 23. Wikipedia Citations Banned at Middlebury. *International Herald Tribune.* http://www-tech.mit.edu/V127/N6/wikipediawire.html (accessed February 26, 2007).

Economist. 2007, March 8. Working the Crowd. http://www.economist.com/science/ tq/displaystory.cfm?story_id=E1_RSGVJJN (accessed March 8, 2007).

Fox, J. 2007, February 15. Getting Rich Off Those Who Work for Free. *Time.* http:// www.time.com/time/magazine/article/0,9171,1590440,00.html (accessed February 16, 2007).

Geist, M. 2007, February 28. Push for Open Access to Research. BBC. http://news.bbc. co.uk/2/hi/technology/6404429.stm (accessed March 1, 2007).

Government of Catalunya. 2005. *Law 16/2005 of 27th December, on Geographical Information and the Cartographic Institute of Catalonia,* J. Crompvoets (English trans.). Institut Cartografic de Catalunya, Barcelona, Spain. http://www.icc.es/web/ content/pdf/ca/common/icc/Llei_Info_geografica_ICC_271205.pdf (accessed March 17, 2007).

Government of Catalunya. 2006. *Decree 398/2006, of the 24th of October, the Development Regulations of Act 16/2005, of the 27th of December,* J. Crompvoets (English trans.). Institut Cartografic de Catalunya, Barcelona, Spain.

Harding, T. 2007, January 21. Google Blots Out Iraq Bases on Internet. *Daily Telegraph* (London). http://www.telegraph.co.uk/news/main.jhtml?xml=/news/ 2007/01/20/wgoogle20.xml (accessed February 18, 2007).

Helft, M. 2007, February 22. Google Challenges Microsoft with New Business Package. *International Herald Tribune.* http://www.iht.com/articles/2007/02/22/ business/google.php (accessed February 26, 2007).

Hussein, A. 2007, February 15. Google Earth, the Survival Tool of War-Torn Iraq. *Daily Telegraph* (London). http://www.telegraph.co.uk/news/main.jhtml?xml=/ news/2007/02/15/wgoogle215.xml (accessed February 18, 2007).

JISC. 2007. European Commission Discusses Future of Scientific Publishing. JISC news item. http://www.jisc.ac.uk/news/stories/2007/02/news_ecconf.aspx (accessed April 20, 2007).

Kablenet. 2007, February 23. DFT to Provide Road Pricing Funding. Kable Government Computing. http://www.kablenet.com/kd.nsf/Frontpage/323A943C 424770808025728B003F31BD?OpenDocument (accessed February 27, 2007).

Kelmelis, J., M. DeMulder, C. Ogrosky, et al. 2003. The National Map: from geography to mapping and back again. *Photogrammetric Engineering and Remote Sensing,* 69: 1109–1118. http://nationalmap.gov/report/PERS_article_forviewing. pdf (accessed April 15, 2007).

Marshall, E. 2000. Scientific misconduct: how prevalent is fraud? That's a million-dollar question. *Science,* 290: 1662–1663. http://www.scienceonline.org/cgi/ content/summary/290/5497/1662 (accessed April 11, 2007).

Millward, D. 2007, February 13. Capital Paid Heavy Price for Congestion Charge. *Daily Telegraph* (London). http://www.telegraph.co.uk/news/main.jhtml?xml=/ news/2007/02/13/nroads113.xml (accessed February 18, 2007).

Reuters. 2007, March 28. Yahoo to Offer E-Mail Storage without End. Reuters. http://www. iht.com/articles/2007/03/28/technology/yahoo.php (accessed March 29, 2007).

Sokal, A. and J. Bricmont. 1999. *Intellectual Impostures.* Profile Books Ltd., London.

Glossary and acronyms

AGI Association for Geographic Information. U.K. national GI association.

AI Artificial intelligence.

ANZLIC Australia and New Zealand Land Information Council.

AOL America Online. Major global Internet service provider.

APPSI U.K. Advisory Panel on Public Sector Information.

APSDI Asia Pacific Spatial Data Infrastructure. A regional SDI initiative promulgated within the United National Regional Cartographic Centers.

ASAP Atypical Signal Analysis and Processing.

ASDI Australian Spatial Data Infrastructure.

BBC U.K. British Broadcasting Corporation. State-owned broadcasting company for radio, television, and Internet services.

CBA Cost–benefit analysis, of which there are many methodologies.

CEA Cost-effectiveness analysis. A form of CBA.

CEN European Committee for Standardization. A European standardization body.

CGDI Canadian Geospatial Data Infrastructure (GeoConnections).

CIO Chief information officer.

Click-use A type of online license permitting users to register once for a resource or resource collection and then use it in future.

Directive An official legal instrument of the European Union, issued jointly by the European Parliament and European Council of Ministers, typically setting out pan-European legislation that must then be enacted across all (27) EU member states.

DNF Digital National Framework. The national GI framework for the U.K.

DOE, DEFRA Department for Environment, Food, and Rural Affairs, U.K. Lead on the U.K. National Spatial Data Infrastructure initiative.

DRM Digital Rights Management. IPR control for digital content.

EC European Commission. The executive body of the European Union.

ECDIS Electronic Chart Display System. Electronic navigation aid.

E-ESDI Environmental European Spatial Data Infrastructure. A regional SDI initiative of the European Commission in 2001–2002; replaced by INSPIRE.

EGII European Geographic Information Infrastructure (now embodied in INSPIRE).

ESA Egyptian Survey Authority; European Space Agency.

EU European Union. The political union of 27 European nations who, by treaty signature, agree to implement harmonized regional legislation.

EULA End-user license agreement.

EUR   Monetary code for the euro.

FEMA   U.S. Federal Emergency Management Agency.

FGDC   U.S. Federal Geographic Data Committee. U.S. authority overseeing the National Spatial Data Infrastructure initiatives.

FOI/FOIA   Freedom of Information (Act).

GAO   U.S. Government Accountability Office; formerly Government Accounting Office.

GDP   Gross domestic product.

GEM   General equilibrium model.

GeoConnections   Canadian SDI.

GeoVMM   Geographic Value Measuring Methodology. A cost–benefit analysis methodology applied to geospatial data.

Geospatial data   Geographic information, spatial data. Any data that contains a location attribute.

GI   Geographic information.

GII   Global information infrastructure.

GIS   Geographic information system.

GOS   Geospatial One-Stop. A U.S. national SDI portal project.

GPS   Global positioning system.

GSDI   Global Spatial Data Infrastructure.

GVA   Gross value added.

HMSO   Her Majesty's Stationery Office in U.K.; now the Office of Public Sector Information (OPSI).

Hoxt   U.S. text messaging service using the Internet.

ICA   International Cartographic Association.

ICT   Information and communications technology.

II   Information infrastructure.

INSPIRE   Infrastructure for Spatial Information in Europe. The pan-European SDI.

IPR   Intellectual property rights. Copyright and patents for GI and GIS.

ISO   International Organization for Standardization.

ITU   International Telecommunication Union.

MAD   Mutually Assured Destruction.

Mash-up   A hybrid application, typically Web based and more common in open-source communities, including GIS.

MCA   MultiCriteria analysis. A form of cost–benefit analysis in which not all costs or benefits need to be assigned monetary values.

Met Office   U.K. national meteorological office. A trading fund.

MetroGIS   Regional GIS system for Minneapolis–St. Paul, MN

MIVC   Management information value chain.

NACo   National Association of Counties (U.S.).

NAP   U.S. National Academies Press.

NCLIS   U.S. National Commission on Libraries and Information Science.

NGDF   National Geospatial Data Framework (U.K.)

NGPO   National Geospatial Programs Office (U.S.)

NHS National Health Service (U.K.)

NIH National Institutes of Health (U.S.)

NII National information infrastructure.

NIMSA National Interest Mapping Services Agreement. Agreement between U.K. government and Ordnance Survey GB to pay for noncommercial activities; agreement ended in March 2006.

NMA National Mapping Agency.

NMCA National Mapping and Cadastral Agency.

NPV Net present value. A metric to measure value of an investment.

NRC U.S. National Research Council.

NSDI National Spatial Data Infrastructure.

NSGIC National States Geographic Information Council (U.S.).

NTIS U.S. National Technical Information Service.

NWS National Weather Service.

ODPM U.K. Office of the Deputy Prime Minister (now abolished).

OECD Organization for Economic Cooperation and Development.

OFT Office of Fair Trading. Anticompetition watchdog agency in U.K.

OGC Open Geospatial Consortium, Inc. International industry-driven interoperability standardization body (not *de jure*).

OGC-E OGC Europe. European division of OGC, Inc.

OMB Office of Management and Budget. Budgetary oversight executive agency of U.S. government.

OPSI U.K. Office of Public Sector Information (formerly HMSO).

OSGB Ordnance Survey of Great Britain. The national mapping agency of England, Wales, and Scotland.

OSNI Ordnance Survey of Northern Ireland. The regional mapping agency for Northern Ireland within the U.K.

PCGIAP Permanent Committee on GIS Infrastructure for Asia and the Pacific. Created by resolution of the United Nations Regional Cartographic Conference for Asia and the Pacific (Beijing, May 1994) and reporting to the UNRCC-AP Conference.

PGIH Public geographic information holder.

PPC Policy process cycle.

PSGI Public sector geographic information. Any GI or spatial data collected, owned, or used by a government agency, at any level of government.

PSI Public sector information. Data collected, owned, or used by a government agency, at any level of government.

RFID Radio frequency identification (chips and associated location technology).

ROI Return on investment. A metric to measure the value of an investment.

RTD Research and Technology Development (EU-funded research program).

SDI Spatial data infrastructure.

Spatial data Any data with a location attribute.

STM   Scientific, technical, and medical information.

TNM   The National Map. U.S. national mapping program.

TOU   Terms of use. Legally binding agreement for software, services, etc.

Trading fund   Form of commercialization under which certain U.K. government agencies operate, mainly to achieve cost recovery for operations.

UKHO   U.K. Hydrographic Office.

UNECA   UN Economic Commission for Africa.

UNECE   UN Economic Commission for Europe (not to be confused with the EC).

UNRCC   United Nations Regional Cartographic Conferences.

UNRCC-AP   United Nations Regional Cartographic Conference for Asia and the Pacific (UNRCC-AP).

USBC   U.S. Bureau of the Census.

USGS   U.S. Geological Survey. The national mapping agency of the U.S.

USPTO   U.S. Patent Office.

VMM   Value measuring methodology. A form of multicriteria analysis used in cost–benefit studies.

VoIP   Voice over Internet Protocol. A way of making phone calls via the Internet.

WIPO   World Intellectual Property Organization.

WIYBY   What's In Your Back Yard? An online information system of the U.K.'s Environment Agency.

Index

Milton Keynes UK
Ingram Content Group UK Ltd.
UKHW040104071024
449327UK00019B/801